Understanding and preventing soil erosion

It is widely recognised that agriculture is a significant contributor to global warming and climate change. Agriculture needs to reduce its environmental impact and adapt to current climate change whilst still feeding a growing population, i.e. become more 'climate-smart'. Burleigh Dodds Science Publishing is playing its part in achieving this by bringing together key research on making the production of the world's most important crops and livestock products more sustainable.

Based on extensive research, our publications specifically target the challenge of climate-smart agriculture. In this way we are using 'smart publishing' to help achieve climate-smart agriculture.

Burleigh Dodds Science Publishing is an independent and innovative publisher delivering high quality customer-focused agricultural science content in both print and online formats for the academic and research communities. Our aim is to build a foundation of knowledge on which researchers can build to meet the challenge of climate-smart agriculture.

For more information about Burleigh Dodds Science Publishing simply call us on +44 (0) 1223 839365, email info@bdspublishing.com or alternatively please visit our website at www.bdspublishing.com.

Related titles:

Advances in Conservation Agriculture Volume 1
Print (ISBN 978-1-78676-264-1); Online (ISBN 978-1-78676-266-5, 978-1-78676-267-2)

Advances in Conservation Agriculture Volume 2
Print (ISBN 978-1-78676-268-9); Online (ISBN 978-1-78676-270-2, 978-1-78676-271-9)

Advances in Conservation Agriculture Volume 3
Print (ISBN 978-1-78676-475-1); Online (ISBN 978-1-78676-477-5, 978-1-78676-478-2)

Advances in measuring soil health
Print (ISBN 978-1-78676-426-3); Online (ISBN 978-1-78676-428-7, 978-1-78676-429-4)

Improving soil health
Print (ISBN 978-1-78676-670-0); Online (ISBN 978-1-78676-672-4, 978-1-78676-673-1)

Managing soil health for sustainable agriculture Volume 1
Print (ISBN 978-1-78676-188-0); Online (ISBN 978-1-78676-190-3, 978-1-78676-191-0)

Managing soil health for sustainable agriculture Volume 2
Print (ISBN 978-1-78676-192-7); Online (ISBN 978-1-78676-194-1, 978-1-78676-195-8)

Chapters are available individually from our online bookshop:
https://shop.bdspublishing.com

climate•SMART•publishing
IN AGRICULTURAL SCIENCE

BURLEIGH DODDS SERIES IN AGRICULTURAL SCIENCE
NUMBER 146

Understanding and preventing soil erosion

Edited by Dr Karl Manuel Seeger, University of Trier, Germany

burleigh dodds
SCIENCE PUBLISHING

Published by Burleigh Dodds Science Publishing Limited
82 High Street, Sawston, Cambridge CB22 3HJ, UK
www.bdspublishing.com

Burleigh Dodds Science Publishing, 1518 Walnut Street, Suite 900, Philadelphia, PA 19102-3406, USA

First published 2024 by Burleigh Dodds Science Publishing Limited

Library of Congress Control Number: 2023952469

British Library Cataloguing in Publication Data
A catalogue record for this book is available from the British Library

ISBN 978-1-80146-379-9 (print)
ISBN 978-1-80146-381-2 (PDF)
ISBN 978-1-80146-380-5 (ePub)
ISSN 2059-6936 (print)
ISSN 2059-6944 (online)

DOI: 10.19103/AS.2023.0131

Typeset by Deanta Global Publishing Services, Dublin, Ireland

Contents

Series list

Title	Series number
Achieving sustainable cultivation of maize - Vol 1 From improved varieties to local applications Edited by: Dr Dave Watson, CGIAR Maize Research Program Manager, CIMMYT, Mexico	001
Achieving sustainable cultivation of maize - Vol 2 Cultivation techniques, pest and disease control Edited by: Dr Dave Watson, CGIAR Maize Research Program Manager, CIMMYT, Mexico	002
Achieving sustainable cultivation of rice - Vol 1 Breeding for higher yield and quality Edited by: Prof. Takuji Sasaki, Tokyo University of Agriculture, Japan	003
Achieving sustainable cultivation of rice - Vol 2 Cultivation, pest and disease management Edited by: Prof. Takuji Sasaki, Tokyo University of Agriculture, Japan	004
Achieving sustainable cultivation of wheat - Vol 1 Breeding, quality traits, pests and diseases Edited by: Prof. Peter Langridge, The University of Adelaide, Australia	005
Achieving sustainable cultivation of wheat - Vol 2 Cultivation techniques Edited by: Prof. Peter Langridge, The University of Adelaide, Australia	006
Achieving sustainable cultivation of tomatoes Edited by: Dr Autar Mattoo, USDA-ARS, USA and Prof. Avtar Handa, Purdue University, USA	007
Achieving sustainable production of milk - Vol 1 Milk composition, genetics and breeding Edited by: Dr Nico van Belzen, International Dairy Federation (IDF), Belgium	008
Achieving sustainable production of milk - Vol 2 Safety, quality and sustainability Edited by: Dr Nico van Belzen, International Dairy Federation (IDF), Belgium	009
Achieving sustainable production of milk - Vol 3 Dairy herd management and welfare Edited by: Prof. John Webster, University of Bristol, UK	010
Ensuring safety and quality in the production of beef - Vol 1 Safety Edited by: Prof. Gary Acuff, Texas A&M University, USA and Prof. James Dickson, Iowa State University, USA	011
Ensuring safety and quality in the production of beef - Vol 2 Quality Edited by: Prof. Michael Dikeman, Kansas State University, USA	012
Achieving sustainable production of poultry meat - Vol 1 Safety, quality and sustainability Edited by: Prof. Steven C. Ricke, University of Arkansas, USA	013
Achieving sustainable production of poultry meat - Vol 2 Breeding and nutrition Edited by: Prof. Todd Applegate, University of Georgia, USA	014

Introduction

It's been suggested that around 12 million hectares of agricultural land are affected by soil erosion each year. If the degradation of the world's soil reservoir continues, many have estimated that this could lead to a 30% reduction in global food production by 2040.

This book provides a comprehensive overview of recent research on this issue. The chapters are split into three parts: Part 1 chapters explore the mechanisms of soil erosion, covering soil erodibility, water-based soil erosion, and the impact of climate change on soil erosion. Chapters in Part 2 describe how soil erosion can be measured and modelled, including proximal instrumental techniques, tracking sediment transport from agricultural soils, and use of the Revised Universal Soil Loss Equation (RUSLE). Finally, Part 3 chapters consider mitigating soil erosion, covering the use of buffer strips, soil stabilisers and zero/no-tillage.

Part 1 Mechanisms

The first chapter of the book focuses on advances in understanding soil erodibility. Chapter 1 opens by discussing approaches to defining soil erodibility. Next, two key soil properties affecting erodibility are reviewed: aggregate stability and shear stress stability. The chapter assesses individual and combined techniques for measuring these properties to gauge the erodibility of a soil, including laboratory and field methods.

Chapter 2 explores soil erosion by water processes, particularly at an agricultural field scale, and current developments in erosion process modelling. The chapter begins with a description of hillslope and field erosion processes, including interrill, rill, and ephemeral gully erosion processes. This is followed by a section highlighting subsurface hydrology and climate input effects on upland erosion. Discussion then moves to methods for estimating and predicting water-based soil erosion, including the Water Erosion Prediction Project (WEPP).

The following chapter assesses the impact of climate change on soil erosion by water. Chapter 3 begins by first describing the influence of climate change on event and annual erosivity, then moves on to discuss the influence of climate change on the protective effect of crops in maintaining soil integrity. A section on snowmelt erosion is also provided, before the chapter concludes with an overview of the multiple influences that climate change has in affecting current and future soil erosion.

Part 2 Measuring soil erosion

Part 2 opens with an overview of the advances in proximal instrumental techniques for measuring gully and rill erosion of agricultural soils. Chapter 4 begins by defining both ephemeral gully erosion and rill erosion, before describing ways to measure them and determine the shape and size of their cross-sections. This is followed by a review of the issues in measuring shape and size, covering the effects of the density of measurement points when estimating cross-sectional areas and of the density of sections, as well as determining the correct width of the gully/rill. Finally, a case study of on site characterisation of small eroded channels is presented.

Chapter 5 discusses advances in tracking sediment transport from agricultural soils. This is critical in both assessing levels of soil erosion and its wider impacts on the surrounding environment. The chapter begins by outlining the difficulties of tracking soil particles due to their size which is typically less than 2mm. It then explores two different approaches: ways of tracking both bulk sediment and tracking individual particles.

The subject of Chapter 6 is advances in modelling soil erosion risk. The chapter opens with an exploration of the development of the universal soil loss equation (USLE) and revised universal soil loss equation (RUSLE and RUSLE2) models. This is followed by sections introducing how these models account for variables such as rainfall, soil erodibility, slope length and gradient, cropping/cover, and area control practice factors. Finally, a case study of using high spatial resolution precipitation data to account for the impact of rainfall in modelling soil erosion risk is presented.

Part 3 Mitigating soil erosion

Chapter 7 reviews the development and effectiveness of establishing buffer strips in preventing sediment and nutrient losses to surface waters. The chapter explains the effects of buffer strips on the capture of sediment and sediment-associated nutrients, such as phosphorus and nitrogen, as well as new methods for improving the effectiveness of buffer strips. The chapter then concludes with case studies reviewing how buffer strips have been implemented and regulated in Denmark, Scotland and Norway.

The following chapter outlines the effects of no-till (NT) and related practices on preventing soil erosion. Chapter 8 starts by focusing on the effects NT practices have on soil properties related to erosion risk, such as bulk density, pore size distribution, saturated hydraulic conductivity, available water holiday capacity, compaction, penetration resistance, crusting/sealing and infiltration, aggregate size, stability and soil structure and also soil organic matter. This is then followed by sections on NT practices to minimise water and wind erosion risk.

The final chapter examines the use of soil stabilisers to prevent erosion. Chapter 9 reviews the potential of three types of soil stabilizing agents: gypsum and gypsiferous materials, synthetic polymers, and organic waste materials. First, the potential of gypsum and gypsiferous materials as soil stabilisers is considered, followed by that of polymers, covering furrow irrigation, rain and overhead irrigation conditions, and structure stability. Finally, the uses of organic residues, such as animal wastes, biosolids, biochar, and food-industry wastes, are explored as soil stabilisers..

Part 1

Mechanisms

Chapter 1

Advances in understanding soil erodibility

Manuel Seeger, University of Trier, Germany

1 Introduction: defining soil erodibility

Soil erosion is a major threat to land productivity and has been the subject of scientific research since the nineteenth century. In his systematic plot experiments, Wollny (1879) and his colleagues identified the most important parameters affecting soil erosion such as vegetation cover and rainfall intensity. Soil erosion is recognised as a highly complex and non-linear process affected by numerous variables in which soil particles are detached from the soil matrix and entrained by flowing water or wind which transport them elsewhere (Duttmann, 2001). Soil erodibility could be simply defined as the vulnerability of a soil to suffer erosion (Porta Casanellas et al., 2011; Stahr et al., 2020). Auerswald et al. (1996) define soil erodibility in terms of soil loss.

Some researchers discuss erodibility in terms of identifying factors affecting vulnerability to erosion such as grain size distribution or humus/organic matter content of soil (Blume et al., 2010, 2016). Other authors define erodibility in terms of resistance to shear stress, for example using an excess shear stress equation including critical shear stress and an erodibility coefficient (Al-Madhhachi et al., 2014). In the LISEM model (Roo et al., 1996), soil erodibility is conceptualised in terms of factors such as aggregate stability and the transport capacity of flowing water. The EROSION-3D Model is derived

http://dx.doi.org/10.19103/AS.2023.0131.01

from factors such as soil texture, land use, crops planted and tillage practices (Schmidt et al., 1999).

Other authors define soil erodibility in terms of the *K*-factor of the Universal Soil Loss Equation (USLE) model (Wishmeier and Smith, 1978) and its later iterations such as Revised USLE (RUSLE), Water Erosion Prediction Program (WEPP) and RUSLE2 (El-Swaify and Dangler, 1977; Alaboz et al., 2021). The *K*-factor represents the rate of soil loss according to a rainfall erosivity index, and is based on a number of variables such as silt, fine sand, clay and organic matter content, an aggregation index and a permeability index (Auerswald et al., 2014). The *K*-factor has for example been used by the European Soil Data Centre to map erodibility risk across the EU (Panagos et al., 2014).

However, since soil erosion is such a complex process (dependent on both intrinsic soil characteristics and a complex range of highly variable external factors) which remains hard to measure or predict, debates about defining and characterising soil erodibility continue (Rickson et al., 2021; Stroosnijder, 2005). This has led to other concepts being proposed such as soil integrity (Sidorchuk, 2005, 2009) or so-called skin factor (Schindewolf and Schmidt, 2012) determined by a rainfall simulator. There have also been attempts to relate soil classification schemes to soil erodibility (Holzhey and Mausbach, 1977; Seeger, 2007). The aim of this chapter is to reflect on different concepts of soil erodibility and its assessment. As recent publications, and recent developments in methods for the assessment of soil erodibility show, there is still a lot to research and debate.

As soil erodibility depends on the susceptibility of aggregates for breakdown and lateral transport, the two main processes involved can be considered to be the breakdown of aggregates and lateral transport by flowing water (Blume et al., 2016). This suggests two main factors determine erodibility of soils:

- Aggregate stability, which is responsible for the supply of smaller, transportable aggregates, their displacement by splash, as well as changing the hydraulic conditions of the soils' surface by eventually reducing infiltration capacity due to crusting (Le Bissonnais and Singer, 1992; Assouline and Mualem, 2001); and
- Shear stress resistance, the susceptibility of soil particles to be entrained and transported by flowing water, which is strongly dependent on particle characteristics like size, shape and bulk density (Le Bissonnais, 1996b; Stahr et al., 2020).

These both are the basis for further characteristics that may strongly influence the erosion of a soil, such as surface roughness, development of flow paths, infiltration changes and many more. They can, however, be regarded as the main measurable factors affecting soil erodibility. These are discussed in more detail below.

2 Factors affecting soil erodibility: aggregate stability

Aggregates are formed when primary soil sand, silt and clay particles bind together. Aggregate stability was introduced as a concept by Schumacher (1864) and discussed in early work on soil science (Ramann, 1905), with their work still providing the foundation for more recent research. As the displacement and transport of soil particles by disruptive forces such as the action of water is largely dependent on particle size (and mass), aggregate stability has often been identified as a key indicator of soil erodibility (Cammeraat and Imeson, 1998; Coote et al., 1988; Le Bissonnais et al., 2007).

The breakdown of soil aggregates is a complex process involving a great variety of physical and chemical mechanisms. Le Bissonnais (1996a) differentiated four main mechanisms of aggregate breakdown:

1. Slaking: this is the breakdown of aggregates by compression of trapped air in the pore system. It depends on the size of the aggregate, its hydraulic conductivity and bulk density, which controls the rate of water uptake (and therefore the entrapment of air) (Roth, 1996). Clay content may have a significant influence on all these mechanisms.
2. Differential swelling: soil minerals, mainly clay minerals and some salts, may swell with different rates and velocities, leading to aggregate breakdown, depending on the mineral and chemical composition of the soil.
3. Physico-chemical dispersion: aggregates and soil particles lose their internal cohesion, leading to total disaggregation.
4. Raindrop impact: the energy of the mass impact on an aggregate leads to its disruption into smaller particles. This mechanism, called splash, is not only responsible for aggregate breakdown, but also the first transport mechanism in soil erosion processes (Kinnell, 1982).

The ability of aggregates to resist breakdown has been linked to factors such as clay content, soil chemistry, time for formation (ageing) and soil microbiology (Almajmaie et al., 2017). Other research has focussed on factors such as aggregate size and pore system characteristics as factors affecting vulnerability to these breakdown mechanisms (Menon et al., 2020; Thomaz et al., 2022).

Soil organic matter (SOM) content has been identified in a number of studies as the most important factor affecting aggregate stability (Barzegar et al., 1997; Basaran et al., 2008; Cantón et al., 2009; Caravaca et al., 2004; Le Bissonnais, 1996b; Le Bissonnais et al., 2007). SOM appears to promote aggregate stability, e.g. by increasing inter-particle cohesion with organo-mineral bonds, as well as increasing hydrophobicity, though a more detailed understanding of how different types of organic matter affect aggregate stability is still lacking (Chenu et al., 2000; Abiven et al., 2009). Basaran et al. (2008), e.g. found a clear correlation between aggregate stability and different types of organic matter. However, Caravaca et al. (2004) found that organic matter content did not affect the stability of macro-aggregates (<250 μm) though it was important for the stability of micro-aggregates (<20μm). Recent research is beginning to clarify the effects of different types of organic matter on aggregate stability (e.g. Sonsri and Watanabe, 2023).

Modern machine learning algorithms are now being applied to develop new pedotransfer functions (PTFs) used to predict aggregate stability from known soil characteristics or from remote sensing data (Bouslihim et al., 2021; Rivera and Bonilla, 2020; Zeraatpisheh et al., 2021). All these studies rely mainly on SOM as a primary indicator of aggregate stability, highlighting both its importance and the need for more research on the role of SOM. To overcome the potential limitations of these simple, almost one-dimensional, attempts to explain aggregate stability, some research groups have looked at the interaction of SOM, minerals and solutes (mainly cations) which affect attractive forces promoting aggregate formation and cohesion (Li et al., 2023; Yu et al., 2017; Yu et al., 2020).

3 Factors affecting soil erodibility: shear stress stability

The resistance of aggregates to the shearing forces of rainfall, wind and flow will affect ease of particle detachment resulting in erosion (Bones et al., 2017). Shear strength can be defined as the resistance to deformation by the action of tangential (shear) stress. Soil shear strength is determined by cohesion between soil particles and friction and interlocking mechanisms which prevent particles sliding over each other. A distinction is made between true and apparent cohesion (Ravindran and Gratchev, 2022):

- True cohesion is a function of soil minerology and results from chemical bonds between particles.
- Apparent cohesion results from water tension in the soil and is strongly influenced by water content.

The minimum amount of shear stress needed to initiate particle movement is known as critical shear stress and is affected by soil properties such as clay and organic matter content, moisture, bulk density, plastic and liquid limit and particle shape, which interact, e.g. with a potential tipping point when soil moisture content reaches the plastic limit (which measures deformation properties of soils) in certain conditions (Knapen et al., 2007; Singh and Thompson, 2016; Zhang et al., 2022). Other factors include management practices such as tillage which affect critical soil properties such as bulk density and water content (Havaee et al., 2015; Hongde et al., 2022; Kang et al., 2021).

4 Assessing soil erodibility: aggregate stability

Le Bissonnais (1996a) provided an extensive review of a range of methods to assess soil aggregate stability relating it to its susceptibility to crusting and erosion. I discuss some of the more common (and, in some cases, de facto standard) types of measurement which address different mechanisms of aggregate breakdown.

4.1 Wet-sieving method

The wet-sieving method of air-dried soil (Elliott, 1986) is one of the most common methods, as it allows the assessment and comparison of very different soil types (Six et al., 2000a; Six et al., 2000b) by sorting aggregates into different size classes. It addresses slaking as a breakdown mechanism in combination with hydraulic and mechanical processes involved in aggregate breakdown (Baumgartl and Horn, 1993). A detailed description of the procedure can be found in Six and Hüppi (2019). Submerging aggregates <8 mm diameter into demineralised water and subsequently sieving them through meshes of 2 mm, 0.25 mm and 0.053 mm width provides a distribution of aggregates according to size and stability. This can be expressed as the mean weight diameter (MWD):

$$\mathrm{MWD} = \sum A_i * P_i,$$

where A_i = mean size of aggregate size class and P_i = proportion of respective aggregate size class.

It is possible to wet-sieve aggregates that are pre-wetted before being submerged. This completely changes the mechanism of aggregate breakdown, as differential swelling or dispersion may now affect the aggregates. Combining them allows a comprehensive assessment of aggregate stability, including soil particle composition, chemical status and porosity within aggregates. Six et al. (2000a) have proposed a normalised stability index (NSI), including a correction for sand. This correction is necessary as sand may falsely increase the values for stable medium and small aggregates. The method is relatively easy to use and allows simultaneous treatment of several samples in standard commercial devices (e.g. Eijkelkamps wet-sieving apparatus).

4.2 Percolation method

The percolation method is not used very widely, although it offers a rapid and easy way to assess aggregate stability, even under field lab conditions. The method was developed by Sekera and Brunner (1943) and enhanced by Becher and Kainz (1983). It uses the same mechanism for an aggregate breakdown as the wet-sieving method described above: slacking of aggregates using forced wetting. Streaming water through the sample adds shear forces to the disaggregation process. In contrast with the wet-sieving method, it uses percolation through a probe to assess the stability of aggregates: a rapid reduction of the percolation rate indicates low aggregate stability (as pores in the sample are clogged by the dispersed material), and vice versa.

Aggregates between 1 mm and 2 mm in size are used. This makes this method susceptible to the amount of sand in the sample, as an increasing sand content may increase apparent aggregate stability. Mbagwu and Auerswald (1999) have proposed a sand-related correction of aggregate stability, which reduces the measured value linearly to the amount of sand in the sample.

4.3 Single-drop-impact count

On a natural, uncovered soil surface the impact of raindrops is considered one of the most important processes in soil erosion (Ellison, 1947; Luk, 1979; Marzen et al., 2015; Marzen and Iserloh, 2021). The process involved combines the effects of slaking and the shear forces exerted by drop impact and the splattering water (Amelung et al., 2018). Despite this, the single-drop-impact-count method (Low, 1967; Smith and Cernuda, 1951) is not as widely used as, e.g. the wet-sieving method but has been used in gathering data for the LISEM model (Roo et al., 1996).

In this method, the number of droplets needed to destroy a soil aggregate is counted and taken as a measure of aggregate stability (Imeson and Vis, 1984). Single aggregates with a size between 3 mm and 5 mm are selected from a soil sample, and bombarded with droplets with a mass of 0.1 g from 1 m height. When the aggregates pass through a 3-mm width mesh, they are considered broken. Analogous to the wet-sieving-method, a pretreatment of the samples by moistening may be also applied. As single aggregates need to be selected and treated with the droplets, the method is laborious, which may be one of the reasons of its relatively low use in the assessment of aggregate stability.

4.4 Comparison and evaluation

In a small case study, we used these three methods to assess aggregate stability in vineyard soils of the Mosel area (Germany) under different treatments in an organically managed vineyard (Seeger et al., 2020). The steep sloping vineyards were characterised by a Eutric Skeletal Regosol (IUSS Working Group WRB, 2015) (one of the internationally recognised soil groups), developed on Devonian slates and characterised by >50 % of coarse fragments (Ries et al., 2022). The fine earth fraction has a sandy loam texture. Underneath the grapevines two types of aromatic herbs (*Origanum vulgare*, *Thymus vulgaris*) were planted to protect the soil against erosion, and to supress the growth of weeds. As a control, mechanical weed suppression was used as a standard method in organic farming. After only 3 years of experiment, several positive effects on soil health were found, with very few negative effects on grape or must production (Dittrich et al., 2021).

A detailed analysis of the aggregate stability of the two treatments and the control was performed by applying the wet-sieving method (slaked and rewetted), the percolation method and the single drop method, also with dry and rewetted aggregates. Despite showing very low correlation coefficients (Spearmans ρ) between the measurements, a qualitative comparison between the treatments showed the same trend (Fig. 1):

- The highest aggregate stability was found with all assessment methods for grapevines combined with *Thymus vulgaris*; and
- Mechanical weed control (control) resulted in the lowest aggregate stability.

Such clear differences from all assessment methods after only after 3 years of the diversification experiment are remarkable. The study also showed the value of using different methods which could then be compared and validated in establishing an overall outcome, despite individual differences in results.

5 Assessing soil erodibility: shear stress stability

Stefanow and Dudziński (2021) review different methods for determining soil shear strength. They conclude that available methods are highly inconsistent

Aggregate stability...	Wet sieving method slaked	Wet sieving method rewetted	Percolation method	Single drop method dry	Single drop method rewetted
Thyme	high	high	high	high*	high
Oregano	lower	lower	lower	lower	lower
Control	lowest*	lowest*	lowest*	lowest	lowest
	* significant difference of averages, $p<0.05$				

Figure 1 Classification of aggregate stability of the different diversification treatments as characterized by different assessment methods.

in their measured values. As examples, I mention one laboratory and one field measurement methodology, which clearly show the challenges in measuring soil shear strength.

5.1 Surface shear stress determination in the laboratory

Shear stress stability is usually described by the Mohr–Coulomb equation. This is applicable on all scales. In the lab, it is possible to measure this with high accuracy on undisturbed samples. With support from Prof. Rainer Horn, the company Eijkelkamp (Royal Eijkelkamp, 2024) developed a device for laboratory determination of surface shear stress with high accuracy (Wojciga et al., 2009). By repeated measurement of the surface resistance against another rough surface with increasing pressure, they were able to determine the Mohr–Coulomb equation for the analysed undisturbed soil samples.

5.2 Surface shear stress determination in the field

Shear resistance is also a scale-dependent property, influenced by various soil and environmental characteristics. This means that field shear resistance may be very different from that measured using undisturbed samples. An automated device simultaneously measuring unidirectional shear resistance and soil surface roughness has been developed for agricultural soils (Schneider et al., 2015; Schneider et al., 2018). The apparatus measures the force needed to cut the soil at variable depths. By characterising the soil surface at high resolution, the authors highlighted the high spatial (and temporal) variability of soil surface shear resistance. They also emphasise that their apparatus is influenced by different processes:

- The cutting of aggregates, and thus measuring aggregate stability;
- The pushing of aggregates and stones or other soil particles to measure shear stress stability; and
- The pulling of roots, which particularly affects the movement of soil particles.

The measurement device gives very detailed information about the soil surface.

6 Assessing soil erodibility: combined measurements

As erodibility is affected by both aggregate stability and lateral transport susceptibility, it is worth investigating methods addressing them both in combination. The jet-stress method is a relatively new method for measuring critical shear stress, as well as the soil erodibility factor, in the field that has been developed in the last decade (Al-Madhhachi et al., 2013; Al-Madhhachi et al., 2011; Ibrahim et al., 2016). It uses a miniaturised laboratory test device applying fluid mechanics principles in soil erosion research (Hanson et al., 1990; Hanson, 1990). Mercier et al. (2012) applied the measurements presented by Hanson and Cook (2004) to solve the basic erosion equation based on the shear stress applied on a soil surface. The measuring principle is based on a water jet that is applied in several bursts on a soil surface, which causes a scour. Depending on the scour dimensions, the critical shear stress as well as the soil erodibility factor can be calculated.

7 Conclusion

The risk of soil erosion, and how to assess the susceptibility of different soils to erosion has been in the focus of agricultural and soil science research since the second half of the nineteenth century. Despite wide used in the literature, soil erodibility has no clear definition in soil science and applied geomorphology. Most research during the last four decades has concentrated on identifying and quantifying soil characteristics that have the greatest influence on aggregate stability as well as the shear stress stability of a soil's surface. In many cases, aggregate stability has been used as a proxy for erodibility.

It is not surprising that most studies point at the soil organic matter content as the most important factor for aggregate stability. Other characteristics, like solutes, mineralogical composition, initial soil moisture, have a minor influence on the dynamics of aggregate breakdown. As the processes of aggregate breakdown are very complex and variable, it is also the assessment of aggregate stability. In fact, most studies have shown that there is no clear direct correlation between individual studies, but that they are almost all valid for comparison purposes and in reaching general conclusions. This problem is even more striking for the assessment of the shear stress resistance of soils. No clear relationship has been established between measured aggregate stability, shear stress resistivity and real-world soil erosion. This is mainly related to the fact that the measurements and the assessment methods focus on a dimension reaching from around 1 mm aggregate size to a few dm of test surface size. However, soil erosion can only be measured from plots and fields starting around 1 m^2. The scales of assessment and observation are very different. We want to explain field size processes with measurements of the aggregate size. There is still little understanding of the interrelations and the interactions of partial processes of erosion across scales. This is still a knowledge gap we need to fill.

8 References

Abiven, S., Menasseri, S. and Chenu, C. 2009. The effects of organic inputs over time on soil aggregate stability: a literature analysis. *Soil Biology and Biochemistry* 41(1), 1–12. doi:10.1016/j.soilbio.2008.09.015 .

Alaboz, P., Dengiz, O., Demir, S. and Şenol, H. 2021. Digital mapping of soil erodibility factors based on decision tree using geostatistical approaches in terrestrial ecosystem. *CATENA* 207, 105634. doi:10.1016/j.catena.2021.105634 . Available at: https://www.sciencedirect.com/science/article/pii/S0341816221004926.

Al-Madhhachi, A. T., Hanson, G. J., Fox, G. A., Tyagi, A. K. and Bulut, R. 2011. Measuring erodibility of cohesive soils using laboratory jet erosion tests. In: *Bearing Knowledge for Sustainability*. World Environmental and Water Resources Congress, Palm Springs, CA. American Society of Civil Engineers, Reston, pp. 2350–2359.

Al-Madhhachi, A. T., Hanson, J. G., , Fox, A. G.,, Tyagi, K. A. , and Bulut, R. 2013. Measuring soil erodibility using a laboratory "mini" JET. *ASABE* 56(3), 901–910. doi:10.13031/trans.56.9742 .

Almajmaie, A., Hardie, M., Acuna, T. and Birch, C. 2017. Evaluation of methods for determining soil aggregate stability. *Soil and Tillage Research* 167, 39–45. doi:10.1016/j.still.2016.11.003 . Available at: https://www.sciencedirect.com/science/article/pii/S0167198716302306.

Amelung, W., Blume, H.-P., Fleige, H., Horn, R., Kandeler, E., Kögel-Knabner, I., Kretzschmar, R., Stahr, K., Wilke, B.-M., Gaiser, T., Gauer, J., Stoppe, N., Thiele-Bruhn, S., Welp, G., Scheffer, F. and Schachtschabel, P. 2018. *Scheffer/Schachtschabel Lehrbuch der Bodenkunde / überarbeitet und ergänzt von Wulf Amelung, Hans-Peter Blume, Heiner Fleige, Rainer Horn, Ellen Kandeler, Ingrid Kögel-Knabner, Ruben Kretzschmar, Karl Stahr und Berndt-Michael Wilke ; mit Beiträgen von Thomas Gaiser, Jürgen Gauer, Nina Stoppe, Sören Thiele-Bruhn und Gerhard Welp ; gegründet von Fritz Scheffer und Paul Schachtschabel, 17.,* überarbeitete *und ergänzte Auflage ed.* Springer Spektrum, Berlin, Germany.

Assouline, S. and Mualem, Y. 2001. Soil seal formation and its effect on infiltration: uniform versus nonuniform seal approximation. *Water Resources Research* 37(2), 297–305. doi:10.1029/2000WR900275 .

Auerswald, K., Fiener, P., Martin, W. and Elhaus, D. 2014. Use and misuse of the K factor equation in soil erosion modeling: an alternative equation for determining USLE nomograph soil erodibility values. *CATENA* 118, 220–225. doi:10.1016/j.catena.2014.01.008 .

Auerswald, K., Kainz, M., Angermüller, S. and Steindl, H. 1996. Influence of exchangeable potassium on soil erodibility. *Soil Use and Management* 12(3), 117–121. Available at: https://bsssjournals.onlinelibrary.wiley.com/doi/epdf/10.1111/j.1475-2743.1996.tb00531.x .

Barzegar, A. R., Nelson, P. N., Oades, J. M. and Rengasamy, P. 1997. Organic matter, sodicity, and clay type: influence on soil aggregation. *Soil Science Society of America Journal* 61(4), 1131–1137. doi:10.2136/sssaj1997.03615995006100040020x .

Basaran, M., Erpul, G. and Ozcan, A. U. 2008. *Variation of Macro-aggregate Stability and Organic Matter Fractions in the Basin of Saraykoy II Irrigation Dam* (vol. 2). Fresenius, Cankiri, Turkey. Environmental Bulletin 17, pp. 224–239. Available at: http://www.scopus.com/inward/record.url?eid=2-s2.0-40649113922&partnerID=40&md5=013691fab806485532b7aeec2251d5a9.

Baumgartl, T. and Horn, R. 1993. Die Bestimmung der Aggregatstabilität - ein Methodenvergleich. *Zeitschrift für Pflanzenernährung und Bodenkunde* 156(5), 385–391. doi:10.1002/jpln.19931560502 .

Becher, H. H. and Kainz, M. 1983. Auswirkungen einer langjahrigen Stallmistdungung auf das Bodengefuge im Lossgebiet bei Straubing. *Zeitschrift für Acker- und Pflanzenbau = Journal of Agronomy and Crop Science* 152, 152–158.

Blume, H. P., Horn, R. and Thiele-Bruhn, S. 2010. *Handbuch des Bodenschutzes 4. Vollständig Überarbeitete Aufl.* WileyVCH Press, Weinheim.

Blume, H.-P., Brümmer, G. W., Fleige, H., Horn, R., Kandeler, E., Kögel-Knabner, I., Kretzschmar, R., Stahr, K., Wilke, B.-M., Scheffer, F. and Schachtschabel, P. 2016. *Soil Science* (1st English edn.). Springer, Heidelberg, New York, Dordrecht, London, 618 pp.

Bones, E. J., Garrow, L. A. and Sturm, T. W. 2017. Incorporating soil erodibility properties into scour risk-assessment tools using HYRISK. *Journal of Infrastructure Systems* 23(3). doi:10.1061/(ASCE)IS.1943-555X.0000340 .

Bouslihim, Y., Rochdi, A. and El Amrani Paaza, N. 2021. Machine learning approaches for the prediction of soil aggregate stability. *Heliyon* 7(3), e06480. doi:10.1016/j.heliyon.2021.e06480 .

Cammeraat, L. H. and Imeson, A. C. 1998. Deriving indicators of soil degradation from soil aggregation studies in southeastern Spain and southern France. *Geomorphology* 23(2–4), 307–321.

Cantón, Y., Solé-Benet, A., Asensio, C., Chamizo, S. and Puigdefábregas, J. 2009. Aggregate stability in range sandy loam soils Relationships with runoff and erosion. *CATENA* 77(3), 192–199. doi:10.1016/j.catena.2008.12.011 . Available at: https://www.sciencedirect.com/science/article/pii/S0341816208001963.

Caravaca, F., Lax, A. and Albaladejo, J. 2004. Aggregate stability and carbon characteristics of particle-size fractions in cultivated and forested soils of semiarid Spain. *Soil and Tillage Research* 78(1), 83–90.

Chenu, C., Le Bissonnais, Y. and Arrouays, D. 2000. Organic matter influence on clay wettability and soil aggregate stability. *Soil Science Society of America Journal* 64(4), 1479–1486. doi:10.2136/sssaj2000.6441479x.

Coote, D. R., Malcolm-McGovern, C. A., Wall, G. J., Dickinson, W. T. and Rudra, R. P. 1988. Seasonal variation of erodibility indices based on shear strengt and aggregate stability in some Ontario soils. *Canadian Journal of Soil Science* 68(2), 405–416. doi:10.4141/cjss88-037 .

Dittrich, F., Iserloh, T., Treseler, C.-H., Hüppi, R., Ogan, S., Seeger, M. and Thiele-Bruhn, S. 2021. Crop diversification in viticulture with aromatic plants: effects of intercropping on grapevine productivity in a steep-slope vineyard in the Mosel Area, Germany. *Agriculture* 11(2), 95. doi:10.3390/agriculture11020095 .

Duttmann, R. 2001. Bodenfeuchte als Steuergrosse der Bodenerosion. *Geographische Rundschau* 53(5), 24–33.

Elliott, E. T. 1986. Aggregate structure and carbon, nitrogen, and phosphorus in native and cultivated soils. *Soil Science Society of America Journal* 50(3), 627–633. doi:10.2136/sssaj1986.03615995005000030017x .

Ellison, W. D. 1947. Soil erosion studies–Part II: soil detachment hazard by raindrop splash. *Agricultural Engineering* 28, 197–201.

El-Swaify, S. A. and Dangler, E. W. 1977. Erodibilities of selected tropical soils in relation to structural and hydrologic parameters. In: Foster, G. R. (Ed.). *Soil Erosion: Prediction and Control. The Proceedings of a National Conference on Soil Erosion May 24 - 26, 1976 Purdue University, West Lafayette, Indiana*. Soil Conservation Society of America, Ankeny, IA, pp. 105–114.

Hanson, G. J. 1990. Surface erodibility of earthen channels at high stresses. Part I - Open channel testing. *Transactions of the ASAE* 33(1), 127–131. doi:10.13031/2013.31305 .

Hanson, G. J. and Cook, K. R. 2004. Apparatur, test procedures, and analytical methods to measure soil erodibility in situ. *Applied Engineering in Agriculture* 20(4). doi:10.13031/2013.16492 .

Hanson, G. J., Robinson, K. M. and Temple, D. M. (Eds.) 1990. *Pressure and Stress Distributions Due to a Submerged Impinging Jet*. American Society of Civil Engineers, Stillwater, USA, pp. 525–530.

Havaee, S., Mosaddeghi, M. R. and Ayoubi, S. 2015. In situ surface shear strength as affected by soil characteristics and land use in calcareous soils of central Iran. *Geoderma* 237–238, 137–148. doi:10.1016/j.geoderma.2014.08.016 . Available at: https://www.sciencedirect.com/science/article/pii/S0016706114003255.

Holzhey, C. S. and Mausbach, M. J. 1977. Using soil taxonomy to estimate K values in the universal soil loss equation. In: Foster, G. R. (Ed.). *Soil Erosion: Prediction and Control. The Proceedings of a National Conference on Soil Erosion May 24 - 26, 1976 Purdue University, West Lafayette, Indiana*. Soil Conservation Society of America, Ankeny, IA, pp. 115–126.

Hongde, W., Dongli, S., Yipeng, Z. and Donghao, M. 2022. Unsaturated soil shear strength of agricultural soils influenced by reclamation sequences in coastal China. *European Journal of Soil Science* 73(3). doi:10.1111/ejss.13237 .

Ibrahim, S. L., Ariffin, J., Abdullah, J. and Muhamad, N. S. 2016. Jet erosion device (Jed)–measurement of soil erodibility coefficients. *Jurnal Teknologi* 78(5–5), 63–67.

Imeson, A. C. and Vis, M. 1984. Assessing soil aggregate stability by water-drop impact and ultrasonic dispersion. *Geoderma* 34(3–4), 185–200. doi:10.1016/0016-7061(84)90038-7 .

IUSS Working Group WRB. 2015. *World Reference Base for Soil Resources 2014, Update 2015: International Soil Classification System for Naming Soils and Creating Legends for Soil Maps*. IUSS Working Group WRB, Rome.

Kang, J., Wei, J., Gan, F. and Li, J. 2021. Shear strength of purple topsoil under different land uses in the three gorges reservoir Area, China. *Mountain Research and Development* 41(3). doi:10.1659/MRD-JOURNAL-D-20-00081.1 .

Kinnell, P. I. A. 1982. Laboratory studies on the effect of drop size on splash erosion. *Journal of Agricultural Engineering Research* 27(5), 431–439. Available at: http://www.scopus.com/inward/record.url?eid=2-s2.0-49049144047&partnerID=40&md5=df24f5030edfafe4e34edaeca9422e1c.

Knapen, A., Poesen, J., Govers, G., Gyssels, G. and Nachtergaele, J. 2007. Resistance of soils to concentrated flow erosion: a review. *Earth-Science Reviews* 80(1–2), 75–109. doi:10.1016/j.earscirev.2006.08.001 . Available at: https://www.sciencedirect.com/science/article/pii/S001282520600136X.

Le Bissonnais, Y. 1996a. Aggregate stability and assessment of soil crustability and erodibility: I. Theory and methodology. *European Journal of Soil Science* 47(4), 425–437. doi:10.1111/j.1365-2389.1996.tb01843.x .

Le Bissonnais, Y. 1996b. Soil characteristics and aggregate stability. In: Agassi, M. (Ed.). *Soil Erosion, Conservation, and Rehabilitation*. CRC Press, Taylor and Francis Group, Boca Raton, pp. 41–60.

Le Bissonnais, Y., Blavet, D., Noni, G. de, Laurent, J. Y., Asseline, J. and Chenu, C. 2007. Erodibility of Mediterranean vineyard soils: relevant aggregate stability methods and significant soil variables. *European Journal of Soil Science* 58(1), 188–195.

Le Bissonnais, Y. L. and Singer, M. J. 1992. Crusting, runoff, and erosion response to soil water content and successive rainfalls. *Soil Science Society of America Journal* 56(6), 1898–1903. doi:10.2136/sssaj1992.03615995005600060042x .

Li, S., Wang, B., Zhang, X., Wang, H., Yi, Y., Huang, X., Gao, X., Zhu, P. and Han, W. 2023. Soil particle aggregation and aggregate stability associated with ion specificity and organic matter content. *Geoderma* 429, 116285. doi:10.1016/j.geoderma.2022.116285 . Available at: https://www.sciencedirect.com/science/article/pii/S0016706122005924.

Low, A. J. 1967. Measurement of stability of moist soil aggregates to falling waterdrops according to Low. In: Low, A. J. (Ed.). *West-European Methods for Soil Structure Determination*. State Faculty of Agricultural Sciences, Gent, 51–78.

Luk, S. H. 1979. Effect of soil properties on erosion by wash and splash. *Earth Surface Processes* 4(3), 241–255. Available at: http://www.scopus.com/inward/record.url?eid=2-s2.0-0018654831&partnerID=40&md5=24bd91ac9decd609bd465afb2851a8b6.

Marzen, M. and Iserloh, T. 2021. Processes of raindrop splash and effects on soil erosion. In: Rodrigo-Comino, J. (Ed.). *Precipitation*. Elsevier, 351–371.

Marzen, M., Iserloh, T., Casper, M. C. and Ries, J. B. 2015. Quantification of particle detachment by rain splash and wind-driven rain splash. *CATENA* 127, 135–141. doi:10.1016/j.catena.2014.12.023 . Available at: http://www.sciencedirect.com/science/article/pii/S0341816214003750.

Mbagwu, J. S. C. and Auerswald, K. 1999. Relationship of percolation stability of soil aggregates to land use, selected properties, structural indices and simulated rainfall erosion. *Soil and Tillage Research* 50(3–4), 197–206. Available at: http://www.scopus.com/inward/record.url?eid=2-s2.0-0033007575&partnerID=40&md5=8f004b759000b7170ea374a6aac8b844.

Menon, M., Mawodza, T., Rabbani, A., Blaud, A., Lair, G. J., Babaei, M., Kercheva, M., Rousseva, S. and Banwart, S. 2020. Pore system characteristics of soil aggregates and their relevance to aggregate stability. *Geoderma* 366, 114259. doi:10.1016/j.geoderma.2020.114259 .

Mercier, F., Bonelli, S., Anselmet, F., Pinettes, P., Courivaud, J. R. and Fry, J. J. 2012. On the numerical modelling of the Jet Erosion Test. In: 6th International Conference on Scour and Erosion. SHF, Paris, France, 8 p.

Panagos, P., Meusburger, K., Ballabio, C., Borrelli, P. and Alewell, C. 2014. Soil erodibility in Europe: a high-resolution dataset based on Lucas. *Science of the Total Environment* 479–480, 189–200. doi:10.1016/j.scitotenv.2014.02.010 . Available at: https://www.sciencedirect.com/science/article/pii/S0048969714001727.

Porta Casanellas, J., López-Acevedo Reguerín, M. and Poch Claret, R. M. 2011. *Introduccion a la Edafologia: Uso y Protección de Suelos* (2nd edn.). Mundi Prensa, Madrid, 535 pp.

Ramann, E. 1905. *Online-Ressource* Bodenkunde, Z. A. (Ed.). Springer, Berlin, Heidelberg.

Ravindran, S. and Gratchev, I. 2022. Effect of water content on apparent cohesion of soils from landslide sites. *Geotechnics* 2(2), 385–394. doi:10.3390/geotechnics2020017 .

Rickson, R. J., Dowdeswell Downey, E., Alegbeleye, G. and Cooper, S. E. 2021. Advances in techniques to assess soil erodibility. In: Otten, W. (Ed.). *Advances in Measuring Soil Health*. Burleigh Dodds Science Publishing, Cambridge, pp. 175–214.

Ries, J. B., Porten, M., Iserloh, T., Marzen, M., Hausener, Y., Dittrich, F. and Seeger, M. 2022. Steillagenweinbau an Mosel und Saar. In: Lorenz, M., Thiele-Bruhn, S. and Ahl, Chr. (Eds). *Exkursionsführer* (vol. 2022), pp. 106–121.

Rivera, J. I. and Bonilla, C. A. 2020. Predicting soil aggregate stability using readily available soil properties and machine learning techniques. *CATENA* 187, 104408. doi:10.1016/j.catena.2019.104408 . Available at: https://www.sciencedirect.com/science/article/pii/S0341816219305508.

Roo, A. P. J. de, Wesseling, C. G. and Ritsema, C. J. 1996. LISEM: a single-event physically based hydrological and soil erosion model for Drainige Basins. I: theory, input and output. *Hydrological Processes* 10(8), 1107–1117. doi:10.1002/(SICI)1099-1085(199608)10:8<1107::AID-HYP415>3.0.CO;2-4 .

Roth, C. H. 1996. Aggregation indices to characterize structural breakdown of dry soil samples by air slaking. *Zeitschrift fur Pflanzenernahrung und Bodenkunde* 159(5), 429–435. Available at: http://www.scopus.com/inward/record.url?eid=2-s2.0-33748957029&partnerID=40&md5=6fc503ab7aaebf23e3cb994ca3ba7e5e.

Eijkelkamp, R. 2024. Surface shear test apparatus. Available at: https://www.royaleijkelkamp.com/products/lab-testing-equipment/soil-physical-research/compression-and-shear-testing/surface-shear-test-apparatus/#section-downloads. Accessed 27 February 2024.

Schindewolf, M. and Schmidt, J. 2012. Parameterization of the EROSION 2D/3D soil erosion model using a small-scale rainfall simulator and upstream runoff simulation. *CATENA* 91(0), 47–55. doi:10.1016/j.catena.2011.01.007 . Available at: http://www.sciencedirect.com/science/article/pii/S0341816211000105.

Schmidt, J., Werner, M.v., Michael, A., 1999. Application of the EROSION 3D model to the CATSOP watershed, The Netherlands. *CATENA* 37 (3–4), 449–456. doi:10.1016/S0341-8162(99)00032-6.

Schneider, R., Thiele-Bruhn, S., Casper, M. C., Raugewitz, K. N., Schardong, A. and Schwab, C. 2015. Räumlich und zeitlich Hoch auflösende Ermittlung der Oberflächenscherfestigkeit und -Rauheit. In: *Räumliche Heterogenität - Erkennen, Abbilden, Validieren oder Ignorieren? Beiträge zum 4. Trierer Workshop zur Niederschlag-Abfluss-Modellierung am 5. und 6. Oktober 2015 in Trier*. Forum für Hydrologie und Wasserbewirtschaftung, Trier. 5.-6.10. DWA, Fachgemeinschaft Hydrologische Wissenschaften, Hennef, pp. 21–26.

Schneider, R., Thiele-Bruhn, S., Schwab, C., Schardong, A., Raugewitz, K. N. and Jackel, M. 2018. High resolution, on-site measurement of soil surface shear strength and roughness with new 4S-device. In: Sychev, V. G. and Müller, L. (Eds). *New methods and results of landscape research in Europe, Central Asia and Siberia. Landscape Monitoring and Modelling*. Publishing House FSBSI «Pryanishnikov Institute of Agrochemistry», Moscow, pp. 197–200.

Schumacher, W. 1864. *Die Physik-I, die Physik des Bodens*. Wiegandt & Hempel, Berlin.

Seeger, M. 2007. Uncertainty of factors determining runoff and erosion processes as quantified by rainfall simulations. *CATENA* 71(1), 56–67. doi:10.1016/j.catena.2006.10.005 . Available at: http://www.sciencedirect.com/science/article/B6VCG-4MH2C7R-4/2/f7161ca6ed6d166f8321e0d0f327453a.

Seeger, M., Dittrich, F., Iserloh, T. and Thiele-Bruhn, S. 2020. Diversifying steep slope viticulture–towards a sustainable intensive agriculture? *Multidisciplinary Digital Publishing Institute Proceedings* 30(1), 51.

Sekera, F. and Brunner, A. 1943. Beiträge zur Methodik der Gareforschung. *Bodenkunde und Pflanzenernährung* 29(1), 169–212. doi:10.1002/jpln.19430290117 .

Sidorchuk, A. 2005. Stochastic modelling of erosion and deposition in cohesive soils. *Hydrological Processes* 19(7), 1399–1417. doi:10.1002/hyp.5568 .

Sidorchuk, A. 2009. A third generation erosion model: the combination of probabilistic and deterministic components. *Geomorphology* 110(1–2), 2–10. doi:10.1016/j.geomorph.2008.12.019 . Available at: http://www.sciencedirect.com/science/article/B6V93-4VK6N40-1/2/43dd7434c33e1805063c284e1a44e66b.

Singh, H. V. and Thompson, A. M. 2016. Effect of antecedent soil moisture content on soil critical shear stress in agricultural watersheds. *Geoderma* 262, 165–173. doi:10.1016/j.geoderma.2015.08.011 . Available at: https://www.sciencedirect.com/science/article/pii/S0016706115300409.

Six, J., Elliott, E. T. and Paustian, K. 2000a. Soil structure and soil organic matter II. A normalized stability index and the effect of mineralogy. *Soil Science Society of America Journal* 64(3), 1042–1049. doi:10.2136/sssaj2000.6431042x .

Six, J. and Hüppi, R. 2019. Aggregate stability and size distribution. In: Álvaro Fuentes, J., Lóczy, D., Thiele-Bruhn, S. and Zornoza Belmonte, R. (Eds). *Handbook of Plant and Soil Analysis for Agricultural Systems. Crop Diversification and Low-Input Farming across Europe: from Practitioners' Engagement and Ecosystems Services to Increased Revenues and Value Chain Organisation*. CRAI, Cartagena, pp. 188–190.

Six, J., Paustian, K., Elliott, E. T. and Combrink, C. 2000b. Soil structure and organic matter I. Distribution of aggregate-size classes and aggregate-associated carbon. *Soil Science Society of America Journal* 64(2), 681–689. doi:10.2136/sssaj2000.642681x .

Smith, R. M. and Cernuda, C. F. 1951. Some applications of water-drop stability testing to tropical soils of Puerto Rico. *Soil Science* 71(5), 337–346. doi:10.1097/00010694-195105000-00002 .

Sonsri, K. and Watanabe, A. 2023. Insights into the formation and stability of soil aggregates in relation to the structural properties of dissolved organic matter from various organic amendments. Special Issue. *Soil and Tillage Research* 232, 105774. doi:10.1016/j.still.2023.105774 . Available at: https://www.sciencedirect.com/science/article/pii/S0167198723001411.

Stahr, K., Kandeler, E., Herrmann, L. and Streck, T. 2020. *Bodenkunde und Standortlehre / Karl Stahr, Ellen Kandeler, Ludger Herrmann, Thilo Streck, 4., vollständig* überarbeitete *Auflage ed*. Verlag Eugen Ulmer, Stuttgart.

Stefanow, D. and Dudziński, P. A. 2021. Soil shear strength determination methods – state of the art. *Special Issue: Soil and Tillage Research* 208, 104881. doi:10.1016/j.still.2020.104881 . Available at: https://www.sciencedirect.com/science/article/pii/S0167198720306632.

Stroosnijder, L. 2005. Measurement of erosion: is it possible? *CATENA* 64(2–3), 162–173. doi:10.1016/j.catena.2005.08.004 . Available at: http://www.sciencedirect.com/science/article/pii/S0341816205001268.

Thomaz, E. L., Araujo-Junior, C. F., Vendrame, R. S. and de Melo, T. R. 2022. Mechanisms of aggregate breakdown in (sub) tropical soils: effects of the hierarchical resistance. *CATENA* 216, 106377. doi:10.1016/j.catena.2022.106377 . Available at: https://www.sciencedirect.com/science/article/pii/S0341816222003630.

Wishmeier, W. H. and Smith, D. D. 1978. *Predicting Rainfall Erosion Losses: A Guide to Conservation Planning*. United States Department of Agriculture/Science and Education Administration. Govt. Printing Office, Washington, DC.

Wojciga, A., Bolte, K., Horn, R., Stepniewski, W. and Bajuk, E. 2009. Surface shear resistance of soil on the micro- to mesoscale. *International Agrophysics* 23(4), 391–398. Available at: https://agro.icm.edu.pl/agro/element/bwmeta1.element.agro-article-f2b4215c-fb73-4164-9b2f-cb328eb698cf.

Wollny, E. (Ed.) 1879. *Forschungen auf dem Gebiete der Agrikultur-Physik*. Carl Winter's Universitätsbuchhandlung, Heidelberg.

Yu, Z., Zhang, J., Zhang, C., Xin, X. and Li, H. 2017. The coupling effects of soil organic matter and particle interaction forces on soil aggregate stability. *Special Issue: Soil and Tillage Research* 174, 251–260. doi:10.1016/j.still.2017.08.004 .

Yu, Z., Zheng, Y., Zhang, J., Zhang, C., Ma, D., Chen, L. and Cai, T. 2020. Importance of soil interparticle forces and organic matter for aggregate stability in a temperate soil and a subtropical soil. *Geoderma* 362, 114088. doi:10.1016/j.geoderma.2019.114088 .

Zeraatpisheh, M., Ayoubi, S., Mirbagheri, Z., Mosaddeghi, M. R. and Xu, M. 2021. Spatial prediction of soil aggregate stability and soil organic carbon in aggregate fractions using machine learning algorithms and environmental variables. *Geoderma Regional* 27, e00440. doi:10.1016/j.geodrs.2021.e00440 . Available at: https://www.sciencedirect.com/science/article/pii/S2352009421000857.

Zhang, J., Shi, D., Jin, H., Li, H., Jiang, N. and Ye, Q. 2022. Characteristics of cultivated layer soil shear strength for sloping farmland in response to soil erosion in the Three Gorges Reservoir Area, China. *CATENA* 215, 106304. doi:10.1016/j.catena.2022.106304 . Available at: https://www.sciencedirect.com/science/article/pii/S0341816222002909.

Chapter 2

Soil erosion by water processes and prediction technology

Dennis C. Flanagan, USDA-ARS National Soil Erosion Research Laboratory, USA

1 Introduction

Soil erosion by water is a natural phenomenon that occurs on upland areas when soil is detached from a surface by impacting raindrops or by flowing water that exerts shear forces. This chapter will describe and discuss various soil erosion by water processes, particularly at agricultural field scale, and also elucidate current erosion process modeling technology and areas for future research.

Multiple types of soil erosion by water can be described, usually with the dominant form changing as the scale of interest increases. These include the following:

1. Raindrop detachment that includes splash erosion and rainfall-enhanced shallow flow transport, all of which are commonly called interrill erosion.
2. Rill detachment that can include headcut formation and advancement, excess flow shear erosion, sediment transport, sediment deposition in rill channels and/or divergent flow areas.

http://dx.doi.org/10.19103/AS.2023.0131.02
Published by Burleigh Dodds Science Publishing Limited, 2024.

3. Ephemeral gully erosion in concentrated flow channels, headcut formation and advancement, excess flow shear stress detachment, sediment transport, and sediment deposition.
4. Larger channel erosion, excess flow shear stress, channel sideslope undercutting, bank side sloughing, sediment transport, and sediment deposition.
5. Classical gully erosion.
6. Mass wasting, such as land slippage and landslides.

In this chapter, the focus will be on agricultural field scales, which typically would include (1)-(3) described earlier, with spatial scale of fields from 1 ha to ~200 ha. As contributing area becomes larger, channels fed by substantial groundwater flows may become perennial streams, and larger erosion features such as classical gullies may form. Steep slopes in areas prone to disturbances such as high rainfall and/or earthquakes may be more prone to mass wasting processes, even when land cover is very high (e.g. forested).

2 Hillslope and field erosion processes

2.1 Raindrop detachment/interrill erosion processes

Raindrops vary in size from 0.5 mm to 6 mm in diameter. Falling from clouds at heights of several thousands of meters, they reach terminal velocities of 2–12 m s^{-1}, depending upon their drop size and atmospheric conditions (Foote and du Toit, 1969). Rainfall kinetic energies can range between 20 and 35 J m_{-2} mm_{-1}, for rainfall intensities between 20 and 220 mm hr_{-1} (Kinnell, 1981).

When the raindrops strike a bare soil surface, their high kinetic energy and momentum compresses a small area and the drop water and surrounding detached soil particles are ejected upwards and outwards from the point of impact (Fig. 1). Soil aggregates can be broken into fragments by these raindrop impacts, and smaller particles (i.e. silt, clay) may move into soil pores with infiltrating water, and plug them. When rainfall rate exceeds infiltration rate, water will pond on the soil surface and then begin to move downhill, initially as shallow sheet flow, carrying detached soil particles with it. Raindrops impacting these shallow water flows will enhance the ability of the flow to transport the sediment. Shock waves created by raindrops impacting very thin water layers also detach and transport larger sediment particles such as coarse sand sizes (Zhang, 2023).

As the shallow water flows move further downslope, they will coalesce into small channels, known as rills. The combination of splash detachment and raindrop-enhanced shallow overland flow sediment transport to rills is commonly known as interrill erosion, or interrill sediment delivery. Meyer et al.

Figure 1 Soil particles being detached by raindrop impact. Raindrops both detach soil and enhance the ability of shallow water flows to transport the already detached particles. Source: Photo credit: USDA-NRCS.

(1975) and Foster and Meyer (1975) first separated upland erosion processes into interrill and rill components.

Numerous researchers have studied splash detachment and interrill erosion over the past 50 years (Young and Wiersma, 1973; Meyer, 1981; Al-Durrah and Bradford, 1982; Guy et al., 1987; Kinnell, 1991; Fox and Bryan, 2000; Zhang and Wang, 2017; Zhang et al., 2020; Kinnell, 2022; Zhang, 2023). Prediction technologies for interrill erosion most often include driving factors such as rainfall intensity and runoff rate, as well as adjustments for slope, cover, etc. (Flanagan and Nearing, 1995; Flanagan et al., 2001).

Control of interrill soil erosion is most often accomplished by increasing the residue and/or vegetative plant cover over the soil. Conservation tillage or no-tillage can leave high levels of crop residues on the soil that protect it from raindrop impact, and prevent soil pores from clogging with dispersed sediments (Carter, 2005). Plant canopies from the growing cash crop or a cover crop also help protect the soil surface from raindrop detachment.

2.2 Flow detachment/rill erosion processes

Rills are defined as small, concentrated flow channels, usually only 4–20 cm in width and 2-10 cm in depth, that on a flat soil surface will form in random locations (Figs 2 and 3), and can be obliterated by tillage operations (Nearing

Figure 2 Rill channels formed on an experimental plot under rainfall simulation near Salisbury, North Carolina. Source: Photo credit: J. E. Gilley, USDA-ARS.

Figure 3 Extensive rill network on a wheat field in the Palouse region of Washington State. Source: Photo credit: D. C. Flanagan, USDA-ARS.

et al., 1997). Once destroyed by tillage, they may reform again in new random locations. In fields cultivated with elevated beds (ridges) and furrows, rills can form in the flow concentrations in the furrows between the ridges.

Detachment is caused by the shear force of the water flow in the rill acting on the rill channel bed and sidewalls. When flow shear stress exceeds the critical shear stress of the soil, rill detachment will begin. Rill detachment capacity is often calculated as the difference between the flow shear stress and the soil critical shear stress, times a rill erodibility parameter (Nearing et al., 1989). Additionally, sediment load that is being transported in the rill flow can act to decrease the ability of the flow to detach more sediment. If the sediment load in the rill flow exceeds the sediment transport capacity of the flow, sediment deposition will occur instead of detachment (Nearing et al., 1989; Foster et al., 1995). Rills may also just transport incoming sediment from above and/or laterally from the sides, with no additional channel detachment or deposition.

Control of rill erosion can also often be accomplished using conservation tillage and cover crops (De Baets et al., 2011) that provide surface residue and roots that strengthen the soil and roughen the surface. Other methods may also be needed, such as terracing to reduce slope length, or soil amendments such as polyacrylamide to strengthen the soil further (Flanagan et al., 2003).

2.3 Flow detachment/ephemeral gully processes

With greater slope lengths and flow accumulations, larger channels can incise into soils. On fields, these will form in locations based on topographical flow convergence (Fig. 4). These channels are typically small enough to be crossed with tillage equipment, and farmers will usually use tillage to cover and fill them in. However, unlike rills, ephemeral gullies will reform in the same locations in a field, due to the water flow convergences.

Figure 4 depicts a very erodible condition in the Loess Hills of southwestern Iowa, where the steep slopes, and bare and loose silt loam soil have experienced severe soil loss from interrill, rill, and ephemeral gully erosion when a large rainfall/runoff event in the spring occurred shortly after seedbed preparation. Most likely in this situation, the farmer would use a field cultivator and/or disk to till the soil, obliterating all of the visible rills and filling in and covering the ephemeral gullies. If another large storm event were to occur, however, the erosion processes would repeat and the gullies would reform in the same locations.

Flow shear stress detachment also is the main physical erosion process occurring in ephemeral gully erosion. The flow will detach the channel bed, and the bed will erode downward. Larger flows of water, however, provide greater detachment capability of the runoff (than in rills). Once the channel bed has been detached downward to a less erodible soil layer, such as a plow pan or

Figure 4 Excessive sheet, rill, and ephemeral gully erosion in a field near Treynor, Iowa. Source: Photo credit: USDA-NRCS.

glacial till subsoil, the downward detachment will usually stop, and the channel will then widen until the flow shear stress acting on the sidewalls is less than the soil critical shear stress. Any subsequent storms will not detach more soil from the channel unless they have greater runoff flows that would widen the channel further. If the farmer tills the surrounding field and fills in the gully channel with loose soil from nearby, the visible ephemeral gully will disappear from view. If another storm event then occurs, the sequence of gully detachment downward, then widening can occur once again.

Control of ephemeral gully erosion usually involves more than conservation tillage or cover crops. Often a grassed waterway is installed where the gully would form, with the waterway channel geometry and vegetation designed to reduce surface flow shear stress to below the critical shear stress of the soil. Additionally, subsurface drainage is often installed below the waterway to lower the water table and decrease the possibility of subsurface lateral flows and seepage. Grade control structures, such as drop boxes, may also be used to safely convey the runoff flows in the waterway to a lower elevation and outlet from the field. Impoundments such as parallel tile outlets or water and sediment control basins may also be used to decrease the channel length and safely convey water out of a field via subsurface tiles.

3 Associated processes affecting soil erosion

3.1 Subsurface hydrology effects on upland erosion

Much of the earliest field erosion research under natural or simulated rainfall was designed in ways that emphasized soils under drained conditions and Hortonian overland flow. USLE plots, for example, were positioned on sloping areas (unit plot was 9% gradient, 22.1 m long, uniform slope) in which only surface runoff and sediment loss were measured, and no accounting of subsurface water flows made. Rainfall simulation experiments in the field have most often been conducted in the summer, when soil conditions are dry and conducive to the equipment and researcher traffic, and also plots are typically uniform shape with only measurement of surface runoff and sediment losses.

However, in reality, fields have a wide range of shapes and slope gradients, and in some cases subsurface hydrology can have substantial effects on the surface runoff, soil detachment, and sediment losses. Huang and Laflen (1996) studied the effects of seepage on soil erosion in a large 5 m long by 1.2 m wide soil box that had water table control and adjustable slope gradients, under rainfall simulation and water inflows. With a Glynwood clay loam soil, they found that sediment concentrations under seepage conditions averaged 22% greater than those under free drainage under simulated rainfall conditions. Sediment concentrations were six times greater at the 5% slope under 20 cm seepage head compared to free drainage (7 days) for the concentrated inflow conditions. They recommended that process-based erosion models should be expanded to predict seepage conditions and their effects on erosion channel head cutting.

Gabbard et al. (1998) utilized the same soil bed and Glynwood clay loam soil as in the Huang and Laflen (1996) study to examine the effects of potential landscape positions (and subsurface hydrologic conditions) on hillslope soil erosion processes. Their results indicated that under drainage conditions interrill processes were dominant, and rill development and detachment were limited. However, rill development and detachment processes were dominant under the seepage conditions, with erosion rates as great as 60 times more under seepage compared to drainage conditions. Thus the seepage conditions may enhance rill and ephemeral gully development on lower backslopes and footslopes, compared to interrill and rill detachment on upper backslopes, and alternative management practices may be needed to control soil loss on landscape positions where seepage occurs.

Wang et al. (2020) and Wang et al. (2021) conducted laboratory flume studies using uniform sand and an Opal clay soil, respectively, to examine subsurface

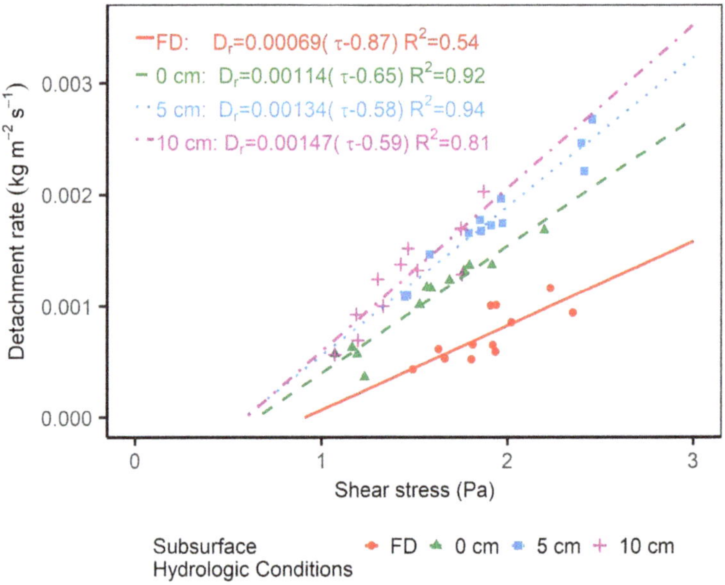

Figure 5 Relationship between detachment rate (D_r) and flow shear stress (τ) under different subsurface hydrologic conditions in experiments of Wang et al. (2021) using an Opal clay soil. Equations show calculated rill erodibilities and critical shear stress values for each subsurface hydrologic condition (FD - free drainage, 0 cm - saturation, 5 cm seepage head, 10 cm seepage head).

hydrologic effects on rill erodibility parameters and sediment transport capacity. Wang et al. (2021) found a definite effect of the subsurface hydrology on rill detachment, and rill erodibility and critical shear stress for the Opal clay soil under free drainage, saturation, 5 cm seepage head, and 10 cm seepage head (Fig. 5). Free drainage resulted in the greatest critical shear stress and smallest rill erodibility, compared to either saturation or seepage conditions. In a similar experiment with uniform sand and sediment inflow provided from a mechanical sediment feeder at the top of the flumes, Wang et al. (2020) determined substantially greater flow sediment transport capacity for the saturation and seepage conditions, compared to the free drainage (Fig. 6).

Further research is needed on subsurface hydrologic effects across a wider range of soils, and how soil detachment and sediment transport processes are affected. Additionally on nonuniform-shaped hillslopes and landscapes, ways to account for changing subsurface hydrologic conditions on erosion processes are needed in new modeling approaches. And, additionally, modeling techniques need to be able to better simulate the effects of subsurface tile drainage practices, and new controlled drainage management systems.

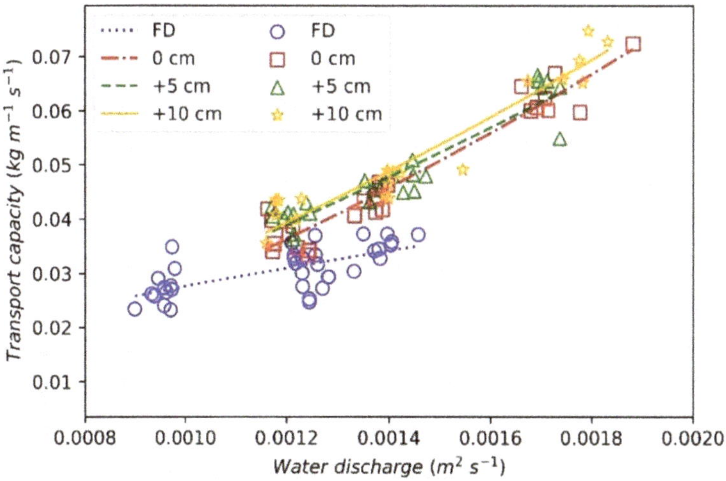

Figure 6 Relationships between sediment transport capacity and water discharge under free drainage, saturation, 5 cm seepage, and 10 cm seepage for uniform sand. Source: Wang et al. (2020).

3.2 Climate input effects on upland erosion

As mentioned earlier, rainfall and/or runoff are the driving factors in interrill, rill, and ephemeral gully erosion processes. Almost all soil erosion prediction tools, both empirical and process-based assume some measure of climate stationarity. For example, the Revised Universal Soil Loss Equation version 2 (RUSLE2) is currently utilizing weather records from 1960 to 1999 in their rainfall/runoff erosivity database. The Water Erosion Prediction Project (WEPP) model has recently updated their climate database to utilize 1974-2013 weather records for development of CLIGEN weather generator parameters (Srivastava et al., 2019) for 2765 stations in the U.S. In both cases, these modeling systems assume that the weather, rainstorms, and runoff events in the future will approximate those in their database records.

However, assumptions of stationarity may no longer be valid, as the effects of climate change insert greater variability and extremes into precipitation events and daily temperatures. McGehee et al. (2022) compared erosivity values determined using updated climate information (1970-2013) to those in the RUSLE2 database, and found generally greater values in the Midwest, south, and northeast U.S., depending upon the storm exclusion procedure used. They also utilized improved gap-filling and spatial interpolation procedures in their research, which allowed for more detailed erosivity mapping, particularly in the western U.S. that also displayed substantial differences with the existing RUSLE2 erosivity map (McGehee et al., 2022).

Many current and historic weather records contain discrete time interval measurements of rainfall depth (e.g. 15 min) that may not be sufficiently fine to describe the rainfall intensity values that drive the runoff and erosion processes. McGehee et al. (2021) and Flanagan et al. (2020) describe the effects of different discrete interval rainfall measurements on rainfall intensity as well as on predictions of runoff and soil loss. Breakpoint precipitation measurements are ideal but are rarely available. More recent data from the National Oceanic and Atmospheric Administration's (NOAA) Automated Surface Observation System (ASOS) 1-min gauge network provides the opportunity to improve determination of climate inputs to soil erosion models in the future.

As we move into the future, modeling options will likely be needed that will allow evaluation of baseline climate and current land management and cropping systems on predicted runoff and soil loss, as well as effects of future climate scenarios on water and sediment losses. And then even more importantly, how will planned soil conservation methods under baseline climate perform under the future climate scenarios?

4 Process-based erosion by water prediction technology

Beginning in the late 1960s and early 1970s, research on erosion prediction technology moved from the empirical basis of the Universal Soil Loss Equation (USLE; Wischmeier and Smith, 1978) that had been formed over the previous forty years, to models that would mathematically simulate the physical processes of soil detachment (Foster and Meyer, 1972). The erosion component of the Chemicals, Runoff, and Erosion from Agricultural Management Systems (CREAMS) model (Knisel, 1980; Knisel and Douglas-Mankin, 2012) was the first to separate upland erosion processes into rill and interrill components (Foster et al., 1980), though it still utilized USLE factors in detachment simulation. CREAMS could also simulate field channels such as ephemeral gullies or grassed waterways, and impoundments.

Other process-based models begun in the 1980s–1990s included WEPP (Nearing et al., 1989; Flanagan and Nearing, 1995; Flanagan et al., 2001), GUEST (Rose et al., 1983, 1997), EUROSEM (Morgan et al., 1998), and LISEM (de Roo et al., 1996). A recent review article provides details on the many erosion models worldwide (Kar et al., 2022). The remainder of this chapter will focus on the Water Erosion Prediction Project (WEPP) soil erosion model.

5 Water Erosion Prediction Project (WEPP)

The Water Erosion Prediction Project (WEPP) modeling project was initiated in 1985 by the United States Department of Agriculture (USDA), with the

USDA - Agricultural Research Service (ARS) taking the lead, and user agencies USDA Soil Conservation Service (SCS), USDA Forest Service (FS), and United States Department of the Interior (USDI) - Bureau of Land Management (BLM) participating as well. Also involved were numerous U.S. and foreign universities and research institutions. See Flanagan et al. (2007) for details on the model development activities and participating scientists.

5.1 WEPP model description

The WEPP model simulates hydrologic and soil erosion processes for individual storm events (in single-storm simulations) as well as a host of associated processes in continuous simulation mode, including tillage and soil consolidation effects, daily water balance, plant growth, residue management and decomposition, and irrigation. WEPP is a distributed parameter model, meaning that on a hillslope profile there can be multiple regions or strips of different soils and cropping/management (these are known as OFEs - overland flow elements). Additionally, when simulating small watersheds channel flow routing and erosion processes are included, and when impoundments are utilized, sedimentation in them is predicted. There can be multiple hillslopes, channels, and impoundments within a WEPP small watershed simulation, all having different spatial characteristics for soils, crops, and management. It is not possible in this brief chapter to provide details on all of the processes simulated, so the reader should see Flanagan and Nearing (1995), and Flanagan et al. (2001) for more information on all of the model components and scientific equations utilized. The text here will focus on the core hydrology and sediment science incorporated into WEPP.

Basically how the model works is that on any day in the simulation, if a rainstorm event, irrigation event, or snowmelt event occurs, the hydrology routines will determine infiltration of water into the soil, and whether or not any surface runoff will occur. If surface runoff is predicted to occur, then the erosion component will determine the interrill detachment and delivery of sediment to rills, and net rill detachment or deposition will be calculated at a minimum of 100 equally spaced points down the hillslope profile, to the end of the slope. Runoff, soil loss, and sediment yield for each storm event will be summed over time, to allow for calculation of average annual, annual, or monthly outputs. Myriad other outputs, both tabular and graphical are also available.

5.2 WEPP hydrology

The driving hydrologic inputs to WEPP are the daily storm precipitation depth (P), storm duration (D), time-to-peak intensity (T_p), and ratio of peak intensity to average intensity (I_p) contained within a climate input file, either created by hand or generated by the CLIGEN weather generator (Nicks et al., 1995).

The WEPP model will internally then disaggregate that information into a double-exponential shaped storm, with the breakpoints determined used in the infiltration and runoff calculations. Alternately, users can also directly input storm breakpoints (as time and cumulative precipitation) in the climate input file (Flanagan and Livingston, 1995). Recently, a new tool called WEPPCLIFF has become available that facilitates creation of WEPP input climate files from observed precipitation data (McGehee et al., 2020).

WEPP infiltration is predicted using a Green-Ampt Mein-Larson equation (Mein and Larson, 1973), modified for unsteady rainfall (Chu, 1978) to determine cumulative infiltration depth with time (Stone et al., 1992). The two stages of infiltration are when the soil surface is not yet covered with water (ponded) and when the surface is ponded. When the soil surface is not ponded, all rainfall infiltrates, while after pondage infiltration occurs independently of the rainfall distribution. The soil surface random roughness and slope gradient are used to determine depressional storage with an equation from Onstad (1984). Rain water will both infiltrate and act to fill the depressional storage, and when the storage is filled additional rainfall excess will generate runoff (Stone et al., 1992). Over a storm event, the cumulative rainfall excess that generates runoff is determined, and is shown as the area under the light blue dashed line in Fig. 7.

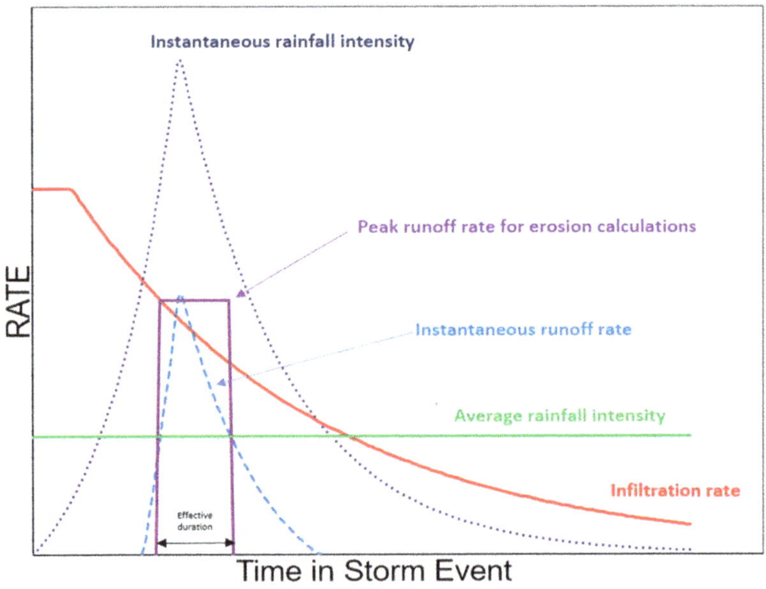

Figure 7 Rainfall, infiltration, and runoff rates within a hypothetical single storm event, as simulated by the WEPP model. The effective runoff duration is the time over which the steady-state soil loss predictions are made, using the peak runoff rate to determine flow shear stress that is used in the rill detachment and sediment transport calculations.

A semi-analytical solution of the kinematic wave model is used to determine the storm runoff hydrograph with runoff rates through time (Stone et al., 1992), or alternately an approximation of the kinematic wave method is used to only determine the storm peak runoff rate (Stone et al., 1995). In Fig. 7, the peak runoff rate is shown as the solid purple line, and the area under that line is fixed to be equal to the same total runoff volume under the actual storm hydrograph (light blue-dashed line). An effective duration of runoff at the peak rate is calculated by dividing the total storm runoff depth by the peak runoff rate. These two variables (peak runoff rate and effective runoff duration) are critical for the subsequent steady-state flow shear stress, sediment transport capacity, and soil loss predictions.

5.3 WEPP hillslope erosion

Solutions to a steady-state continuity equation are used within the WEPP hillslope erosion model to determine sediment load at 100 equally spaced points down a slope profile (or down each OFE on a slope profile). The equation is:

$$dG/dx = D_r + D_i \quad (1)$$

where G is the sediment load (kg s^{-1} m^{-1}), x is distance downslope (m), D_r is rill detachment (or deposition) rate (kg s^{-1} m^{-2}), and D_i is interrill sediment delivery rate to the rills (kg s^{-1} m^{-2}) (Nearing et al., 1989; Foster et al., 1995).

Interrill detachment and sediment delivery (D_i) in Equation 1 is basically determined as the product of an effective rainfall intensity (I_{eff}) and interrill runoff rate (σ_{ir}) times an interrill erodibility (K_i). However the actual full equation used on a daily basis in the WEPP model is:

$$D_i = K_{iadj}\, I_{eff}\, \sigma_{ir}\, SDR_{RR}\, F_{nozzle}\, (R_{sp}/w_r) \quad (2)$$

where K_{iadj} is adjusted interrill erodibility (kg s m^{-4}), I_{eff} is storm event effective rainfall intensity (m s^{-1}), σ_{ir} is storm event interrill runoff rate (m s^{-1}), SDR_{RR} is the interrill storm sediment delivery ratio, F_{nozzle} is an adjustment factor that accounts for sprinkler irrigation nozzle impact energies, R_{sp} is the rill spacing (m), and w_r is the rill width (m) (Foster et al., 1995).

A baseline interrill erodibility (that represents a freshly tilled soil surface with no residue or roots present) is a model input, or can be estimated from a set of default parameter equations if the user enters a zero (0.0) value in the soil input file for baseline K_i (Flanagan and Livingston, 1995). On each day of a model simulation, the baseline interrill erodibility is multiplied by a number of adjustment factors that account for canopy cover, ground cover, roots, slope gradient, consolidation, sealing and crusting, and freezing and thawing effects (Alberts et al., 1995), to determine the K_{iadj} value.

The effective rainfall intensity is estimated in the hydrology component of the model during the storm's infiltration and runoff calculations with:

$$I_{eff} = \frac{1}{t_{eff}} \int I dt \tag{3}$$

where I_{eff} is the effective rainfall intensity (m s^{-1}), I is the breakpoint rainfall intensity (m s^{-1}), t is time in seconds, t_{eff} is the total time over which the rainfall rate exceeds the infiltration rate, and the integral is evaluated over time t_{eff}.

The WEPP model does not currently perform a comprehensive evaluation of splash detachment, enhanced shallow flow sediment transport by raindrop impacts, or sediment deposition on interrill areas. Instead, it uses an interrill sediment delivery ratio (SDR_{RR}) that is a function of the soil surface random roughness, row side-slope, and interrill sediment particle size distribution. Please see Foster et al. (1995) and Flanagan and Nearing (2000) for complete details on the procedure.

The nozzle impact energy factor (F_{nozzle}) was included to account for simulations with water applications from sprinkler irrigation, where nozzle water drop energies may be substantially different (usually less) than those of natural rainfall. The default value of this factor is one (1.0) for natural rainfall.

Rill channels in the WEPP model are assumed to be rectangular in shape, and run from the top to the bottom of a hillslope profile (or each OFE), and occur at regular intervals. Users have the option of direct input of the rill width, spacing, and permanence (permanent or temporary, which means destroyed by tillage operations) in the cropping/management input file. Or by default, the model will assume an initial rill width of zero (0.0) with a spacing of one meter (1.0 m) and temporary. Gilley et al. (1990) found rills to form on average at about 1.0 m spacings on cropland soils at 11 sites (Fig. 2). If users wish to simulate broad sheet flow conditions for an OFE or hillslope, they can set rill width equal to rill spacing and make the setting permanent. With those settings, most if not all detachment would be by interrill processes. However, normally most simulations are made with the default settings, since most users do not have information on rill channel geometry and spacing. During a runoff event, if rill width has not been made permanent, a value for the width will be calculated at the end of each OFE using an equation from Gilley and Weltz (1995):

$$w_{rill} = 1.13\,(Q_{rill})^{0.303} \tag{4}$$

where Q_{rill} is the flow discharge rate (m^3 s^{-1}).

Soil can either be detached from rill channel bottoms and sides, or sediment already in the water flow can deposit to the bed of the channel. The actual process which occurs (detachment or deposition) varies spatially down a slope profile, and is dependent upon the individual storm runoff flow characteristics, flow shear stress and sediment transport capacity at each point down the hillslope, the adjusted erodibility parameters for that day, and the incoming sediment load from the interrill areas and the rill channel above the given point.

Rill detachment occurs when the flow shear stress (τ) exceeds the critical shear stress of the soil (τ_c), and the flow sediment transport capacity (T_c) is greater than the sediment load in the flow (G) at a given point. The equation used is:

$$D_r = k_{radj} (\tau - \tau_{cadj}) [1 - G/T_c] \tag{5}$$

where D_r is the rill detachment rate (kg s^{-1} m^{-2}), k_{radj} is the adjusted rill erodibility for the runoff event (s m^{-1}), τ is the flow shear stress (Pa), τ_{cadj} is the adjusted critical shear stress of the soil for the runoff event (Pa), and T_c is the sediment transport capacity of the flow (kg s^{-1} m^{-1}). Baseline rill erodibility and critical shear stress are model inputs in the soil input file, and represent values for a freshly tilled soil condition with no residue or roots present. If users do not know their values, they can utilize default parameterization equations (Alberts et al., 1995; Flanagan and Livingston, 1995) by inputting zeroes (0.0) in the soil input file for these parameters. On each day of a model simulation, adjustments are made to the rill erodibility for incorporated residue, roots, sealing and crusting, and freezing/thawing, while for critical shear stress adjustments are made for random roughness, sealing and crusting, and freezing/thawing (Alberts et al., 1995).

Elliot and Flanagan (2023) recently published improved estimation equations for baseline K_i, K_r, and τ_c, values by re-analyzing the WEPP cropland field experiment data. These new equations may provide better erosion predictions for users that have access to all of the needed equation parameters.

The '[1 - G/T_c]' term in Equation 5 is a sediment feedback relationship, which limits the detachment capacity of the rill flow when detached sediment is already present. For clear water inflow (or at the top of a rill), $G = 0$, and rill detachment rate will be at a maximum. For another location in a rill where inflowing sediment from the upper channel reach and sediment supplied laterally from the interrill areas is large, detachment rate in the rill will decrease as G approaches T_c in Equation 5.

Sediment transport capacity (T_c) is computed in WEPP as a simplified function of shear stress raised to the 3/2 power times a coefficient. The full Yalin (1963) sediment transport equation is applied at the end of the slope profile, in order to determine the coefficient value (Finkner et al., 1989). The simplified equation is:

$$T_c = k_t \tau_f^{1.5} \tag{6}$$

where k_t is the transport coefficient ($m^{0.5}\ s^2\ kg^{-0.5}$), and τ_f is the flow shear stress that acts on the soil (Pa). For very sandy soils, an additional adjustment to transport capacity is made; see Foster et al. (1995) for more details.

As mentioned earlier, rill width is either input by the user or determined using Equation 4 for a runoff event if the flow discharge for the event exceeds an earlier rate. Flow depth in the rills is determined using an iterative solution of the uniform flow equation, taking into account the roughness of the rill soil surface, crop residue, and live plants that contribute to the total friction factor (Gilley and Weltz, 1995).

With the computed flow depth and rectangular rill channel dimensions, the flow shear stress acting on the soil is then computed using:

$$\tau_f = \gamma R \sin(\arctan S_{rill}) (f_s/f_{tot}) \tag{7}$$

where γ is the specific weight of water ($kg\ m^{-2}\ s^{-2}$), R is hydraulic radius of the water flow in the rill (m), S_{rill} is the rill channel slope gradient ($m\ m^{-1}$), f_s is the Darcy-Weisbach friction factor for the soil, and f_{tot} is the total Darcy-Weisbach friction factor for the rill that includes the soil roughness, residue roughness, and live plant roughness in the rill (Gilley and Weltz, 1995).

Nondimensional equations are utilized within the WEPP model code to determine the solution to the sediment continuity expression at the 100 equally spaced points down a hill (or OFE). The normalizations are described in detail in Nearing et al. (1989), and allow shear stress and sediment transport capacity to be written as polynomial functions of distance down the slope profile that provide easier solution and are more computationally efficient. When in a detachment region and moving from a point with a previously determined sediment load (G) to the next point downslope, the nondimensional version of Equation 1 is solved with a Runge–Kutta numerical method (Nearing et al., 1989; Foster et al., 1995).

Sediment deposition in a rill is predicted when the sediment transport capacity (T_c) is less than the sediment load (G), and is often caused by decreases in T_c due to decreasing slope gradients (e.g. concave slope) or sudden increases in roughness which decrease the flow shear stress acting on the soil (e.g. flow entering a grass filter strip). The equation for deposition in WEPP is:

$$D_r = \frac{\beta v_{eff}}{q}\left[T_c - G\right] \tag{8}$$

where D_r is the rill deposition rate, β is a raindrop-induced turbulence factor, v_{eff} is the effective particle fall velocity ($m\ s^{-1}$), and q is the flow discharge rate per unit width ($m^2\ s^{-1}$). The β factor is currently assigned a value of 0.5 for rainfall-impacted rill flows, and a value of 1.0 for snow melt or furrow irrigation flows (Foster et al., 1995; Flanagan et al., 2001). For depositional regions on a hillslope profile, a closed-form solution of the nondimensional sediment continuity equation is used to determine sediment load at the points within the region. WEPP also computes the sediment particle size distribution for flows entering and leaving detachment and depositional region, using conservation of mass principles - see Flanagan and Nearing (2000) for full details.

5.4 WEPP watershed erosion

In addition to the capability to simulate soil detachment, sediment transport, sediment deposition, and sediment yield from individual storm events on hillslope profiles, the WEPP model also can be used to simulate small field watersheds composed of multiple hillslopes, channels, and impoundments. WEPP watershed simulations are a combination of process-based components and physically-based empirical relationships; the primary components and modeling approaches are as follows:

1. Ascough et al. (1995, 1997) describe the channel hydrology and water balance, including infiltration, percolation, and evapotranspiration that are the same as in the hillslope component. Options to calculate peak runoff rates for channels include a modified rational equation, and an equation from the CREAMS model (Knisel, 1980).
2. Watershed sediment yield results from interrill and rill sediment from hillslope profile areas delivered to channels such as ephemeral gullies, terrace channels, or grassed waterways. Numerical solution of the steady-state spatially varied flow equation is used to determine the flow depths in the channel, and flow shear stress along a channel section. Detachment, transport, and deposition of sediment is based on a steady-state sediment continuity equation, similar to the one used on hillslope profiles (Ascough et al., 1997).
3. Impoundments such as culverts, filter fences, straw bales, and farm ponds can be simulated at outlet points of channels, and will induce deposition of sediment and reduce sediment reaching a watershed outlet. Lindley et al. (1995, 1998a,b) describe the equations and procedures. A continuously stirred reactor is assumed, with adjustments for impoundment shape, stratification, and nonhomogeneous sediment concentrations. The model tracks the volume of deposited sediment in an impoundment, and a mass balance equation is used to predict the outflow sediment concentration (if any).

This chapter's brevity prevents more detailed description of most watershed processes, so the reader is referred to the Ascough et al. (1995, 1997) and

Lindley et al. (1995, 1998a,b) references for full details. The channel erosion calculations will be described briefly hereafter.

A quasi-steady-state sediment continuity equation is used to describe the erosion process in channels during a WEPP watershed simulation. The equation is:

$$dG_{ch}/dx_{ch} = D_F + D_L \tag{9}$$

where G_{ch} is the channel sediment load (kg s^{-1} m^{-1}), x_{ch} is the channel downslope distance (m), D_F is the channel detachment (or deposition) rate (kg s^{-1} m^{-2}), and D_L is lateral sediment inflow from adjacent hillslopes (kg s^{-1} m^{-2}) (Ascough et al., 1997). In the channel erosion calculations, all calculations for sediment load down a channel section are performed for each particle size class, and the same five particle types are used as in the hillslope component: primary clay, primary silt, small aggregate, large aggregate, and primary sand.

Channels can erode downward until a non-erodible layer is reached, with the downward erosion rate described by:

$$E_{ch} = w_{ch} K_{ch}(\tau_{ave} - \tau_{cr}) \tag{10}$$

where E_{ch} is the soil loss per unit channel length (kg m^{-1} s^{-1}), w_{ch} is the channel width (m), τ_{ave} is the average shear stress of the channel flow acting on the soil (Pa), and τ_{cr} is the critical shear stress of the soil in the channel (Pa). The depth of the channel is tracked, and once the depth is equal to the non-erodible layer depth, downward progression stops and the channel will widen until the flow shear stress on the sidewalls is equal to the critical shear stress of the soil. See Foster et al. (1980) for a more complete description of channel downcutting and widening. Any subsequent storm event with less flow will not cause any further detachment in the channel. If a larger storm event with more flow occurs, the channel will widen further. If the field is tilled over and the channel filled with loose soil material again, then the model will reset and another gully can form in a subsequent storm runoff event in the same location.

6 WEPP model interfaces and applications

The WEPP science model is programmed in the Fortran language, and maintained and developed at the USDA-ARS National Soil Erosion Research Laboratory (NSERL) in West Lafayette, Indiana. A variety of interface programs have been developed over the past 30 years by both ARS and others to allow for its easy application, and in many cases targeted toward certain applications. When first publicly released in 1995, the model was provided along with a DOS-based interface; a Windows interface, however, was developed and released in 1999 (Flanagan et al., 2007). The Windows program allows for easy creation, editing, and manipulation of hillslope profiles as well as small field watersheds, with complete user control over all of the WEPP climate, soil, slope, cropping/management, channel, and impoundment inputs. An example hillslope profile in the Windows interface is depicted in Fig. 8, in which a 100-year simulation has been conducted for a slope profile near Coshocton, Ohio with an upper OFE of continuous corn under fall moldboard plow tillage, and a lower OFE with continuous grass. In addition to numerous text outputs related to the runoff, soil loss, and sediment delivery values for the simulation period of interest, and graphical outputs of over 90 variables through the simulation period, there is also the capability to obtain the return period values for daily precipitation, runoff depth, peak runoff rate, and sediment leaving the profile (Fig. 8). The WEPP model, Windows interface, and documentation are available at https://www.ars.usda.gov/Research/docs.htm?docid=10621.

A geospatial interface to WEPP, known as GeoWEPP, is also available as an ArcGIS extension that runs in tandem with the Windows interface program (Renschler, 2003; Flanagan et al., 2013). GeoWEPP is also developed and maintained at the USDA-ARS NSERL, and is available at https://fargo.nserl.purdue.edu/geowepp/index.html. GeoWEPP has recently been applied in a nested, paired watershed study for short- and long-term runoff and sediment yield predictions at watershed outlets (Renschler and Zhang, 2020), and in soil redistribution research (Fig. 9) with Cesium-137 (^{137}Cs) radionuclides (Zhang et al., 2021).

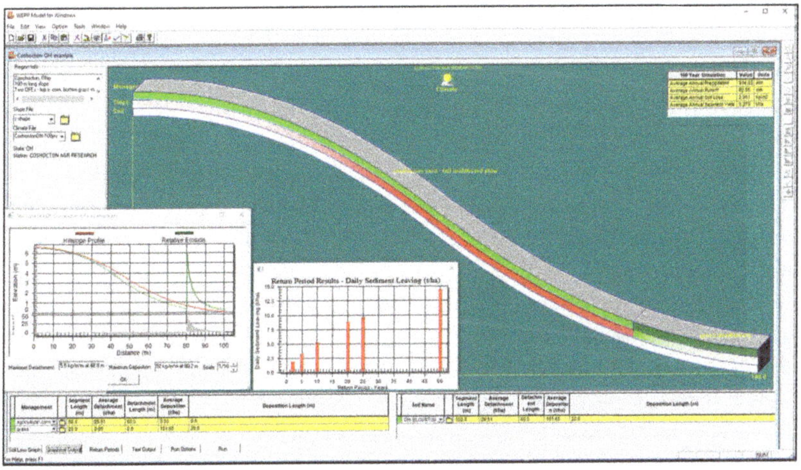

Figure 8 Screen capture of the WEPP Windows interface with a hillslope profile simulation shown for a profile with an upper cropped region and a lower grass filter strip region. Output graphics depict soil detachment and deposition with distance downslope, and return period values of 1–50 years for sediment leaving this profile.

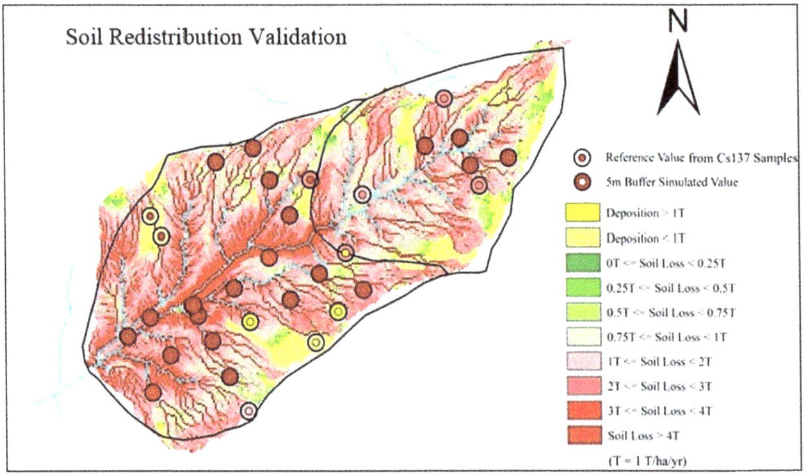

Figure 9 GeoWEPP spatial soil loss output from the study of Zhang et al. (2021) for two nested subcatchments within the Lucky Hills study area on the semi-arid USDA-ARS Walnut Gulch Experimental Watershed (WGEW) near Tombstone, Arizona. Model results were validated using Cesium-137 field data.

More recent WEPP model interfaces have been developed as internet web-based applications, both by USDA-ARS NSERL (https://milford.nserl.purdue.edu), and by other cooperators. The USDA Forest Service (FS) has an entire suite of web interfaces, targeted toward specific needs and users in forested regions, and these can be accessed at https://forest.moscowfsl.wsu.edu/fswepp/ (Elliot, 2004, 2013).

The University of Idaho, in cooperation with the FS and ARS has developed WEPPcloud, a powerful online hydrologic and erosion prediction tool, originally targeted toward forested watersheds in the western U.S., to assess impacts of wildfires on soil loss, by uploading burn severity maps (Lew et al., 2022; Dobre et al., 2022) while setting up the model simulations and inputs. WEPPcloud can be applied anywhere in the U.S. to both disturbed and undisturbed conditions, and there are also options for Europe and Australia (https://wepp.cloud/weppcloud/).

Iowa State University and other cooperators, including USDA-ARS, have developed the Iowa Daily Erosion Project (IDEP; Cruse et al., 2006) and subsequently the Daily Erosion Project (DEP; Gelder et al., 2018) that utilize NexRad radar precipitation data, remotely sensed cropping and management information, NRCS SSURGO soils data, and the WEPP model to determine near real-time estimates of runoff and soil loss from Iowa and surrounding states. The tool predicts soil loss with the WEPP model at ~240 000 randomly sampled locations in Iowa, and a similar density of sampling in the other states covered (Nebraska, Minnesota, and parts of Illinois, Missouri, Kansas, and Wisconsin)

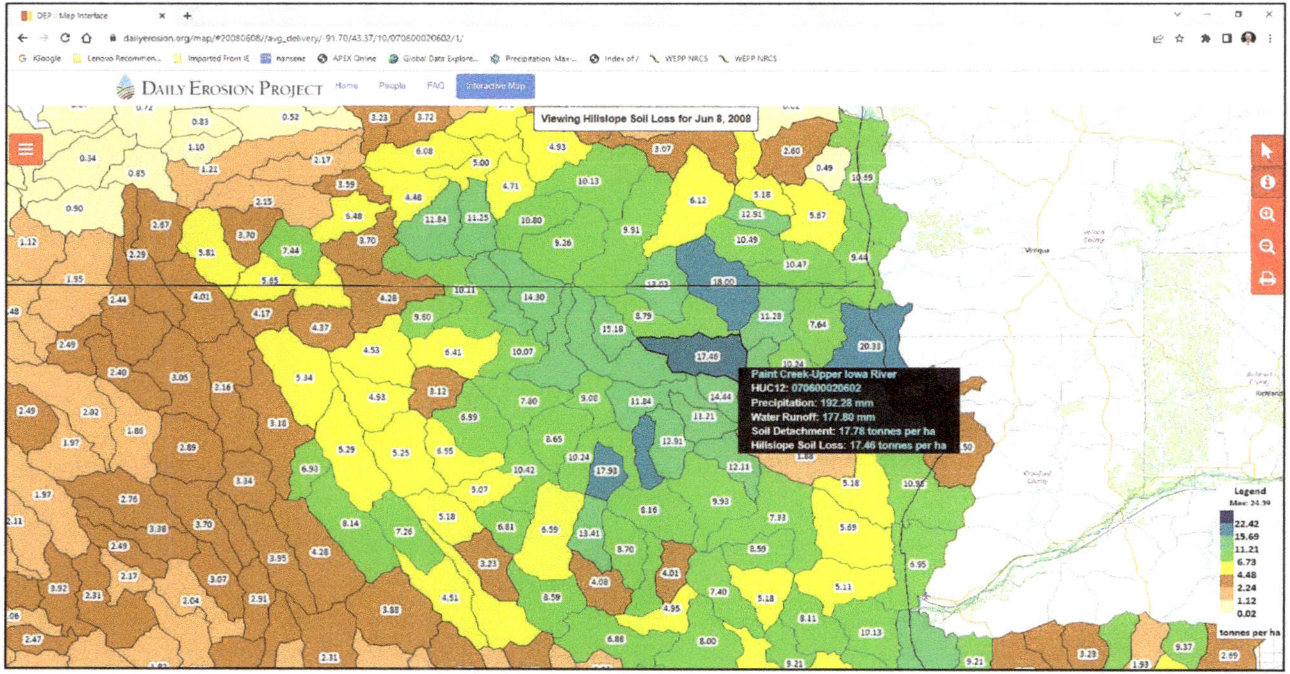

Figure 10 Screen capture of the Daily Erosion Project web interface, with soil loss estimates for June 8, 2008 displayed. Details for each HUC-12 simulation region are provided in the panel that appears wherever the mouse cursor is hovering over. Here the results for the Paint Creek – Upper Iowa River HUC-12 are shown, with an estimated soil loss of 17.46 t/ha for this 192 mm rainfall event. Estimated average event hillslope soil loss for each HUC-12 is also displayed here as the numbers within each catchment.

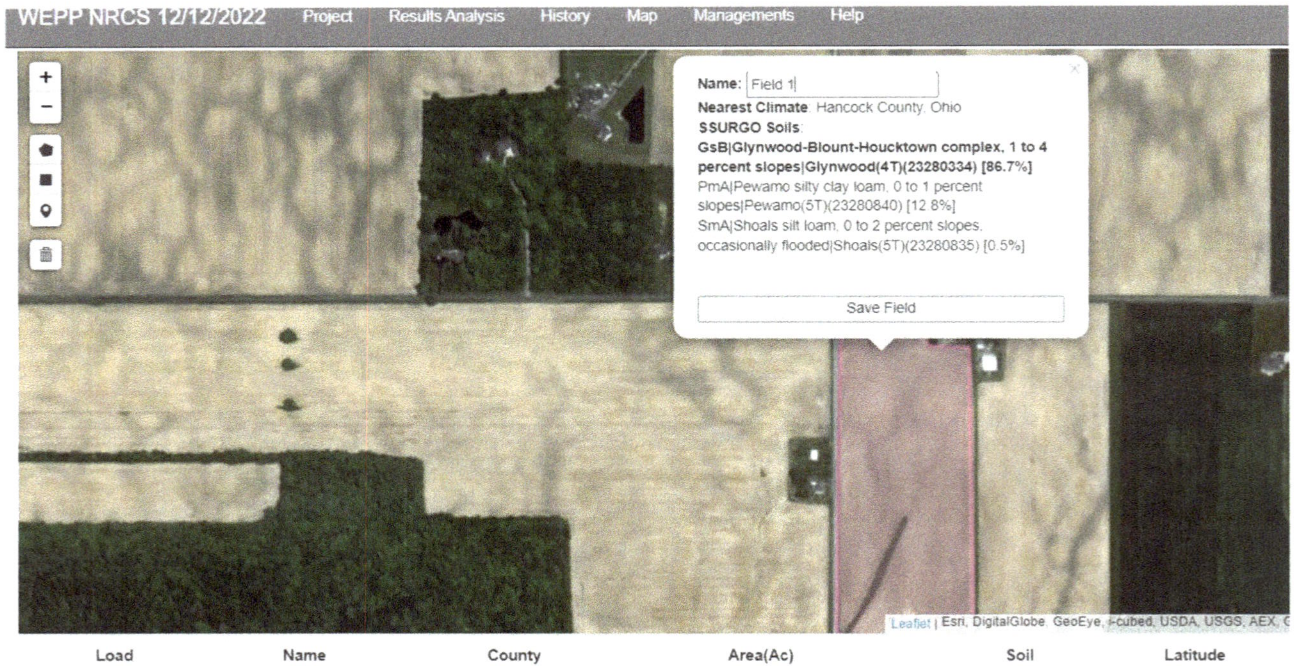

Figure 11 Screen capture of the map screen of the WEPP NRCS web interface for hillslope profile simulations. Here, a field polygon has been outlined (pink color), and the geospatial information including the nearest climate location and SSURGO soil types are noted in the pop-up window. A user can save this field, and those attributes loaded into the setup for a model simulation. https://brenton.nserl.purdue.edu/wepp/.

each day. The average hillslope runoff and soil loss for each HUC12 (Hydrologic Unit Code-12) watershed in the DEP coverage area are then reported in the web-based interface (Fig. 10), and there are numerous options for viewing events or time periods since 2007. Figure 10 shows the largest daily soil loss event predicted for a location in northeast Iowa. The DEP online software is available at https://www.dailyerosion.org/.

The USDA-ARS NSERL has been working with the USDA Natural Resources Conservation Service (NRCS) since 2013 on development of new web-based interfaces for NRCS field office use in soil conservation planning (Flanagan et al., 2023). Interfaces for both hillslope profile simulations (Fig. 11) and field/watershed simulations have been developed, with the hillslope profile interface nearing deployment. NRCS, ARS, and university project staff have also spent considerable time on creating a new Conservation Resources Land Management Operations Database (CRLMOD) for use with these interfaces that contains all of the NRCS required crops, tillage operations, residue management operations, etc. (Kucera and Coreil, 2023). Cooperative work between NRCS, ARS, Colorado State University, Purdue University, and the University of Idaho continues to complete the field/watershed interface (David et al., 2023) and databases by the end of 2024.

7 Current and future WEPP model development

In addition to the current efforts on WEPP model interface and database development with NRCS, there are some other areas of active work. These include WEPP-water quality (WEPP-WQ) model development, WEPP-CO_2 model development, tile drainage component improvement, and future climate scenario applications and testing.

Recently, McGehee et al. (2023) completed incorporation of water quality routines from the SWAT model into multiple OFE and watershed WEPP model simulation capabilities. This built upon earlier work by Savabi et al. (2011) and Wang et al. (2017). While functional, additional work is needed to incorporate the new model code within the WEPP public release software, as well as provide updated interfaces and databases that will allow users to easily apply the new routines for nutrient and pesticide losses.

Srivastava et al. (2023) reported results from a preliminary study where a modified version of the WEPP model was used in which the Penman-Monteith method (Monteith, 1965; Allen et al., 1989) with a stomatal resistance parameter for transpiration calculation was implemented compared to the Penman equation in the normal WEPP model version. The Stockle et al. (1992) equations were also implemented that modify plant growth and biomass production as a function of atmospheric CO_2 level. They ran model simulations with historical baseline climate (1995-2005) and a CO_2 level of 370 ppm, the baseline climate with a CO_2 level of 550 ppm to evaluate effects of only increased CO_2 on biomass production and crop yields, and a future climate (2045–55; GFDL-ESG2G GCM under RCP8.5) with a CO_2 level of 550 ppm to evaluate both increased CO_2 and future climate effects on biomass production and crop yields (Srivastava et al., 2023). They noted that while the model response was mostly as expected, further research is needed with a range of GCM inputs from CMIP5 and CMIP6, to explore WEPP-CO_2 model response from crops under both irrigated and non-irrigated conditions.

Revuelta-Acosta et al. (2021) developed improved subsurface tile drainage routines for the WEPP model that allow it to better estimate tile flows and surface and subsurface runoff from agricultural fields. Additional work is underway to further test these routines, as well as improve performance during winter periods when the soil is frozen and/or thawing, and the WEPP daily water balance changes to account for frost and thaw in the soil. Also, new management options for tile drainage systems such as controlled drainage are needed in the model code and in the user interfaces.

The WEPP model can be applied in modeling studies examining the effects of future climate change scenarios on predicted runoff and soil loss, as well as the effectiveness of soil conservation practices in the future. Flanagan et al. (2014) and Flanagan et al. (2017) provide some examples of WEPP model applications that examine future climate impacts. Trotochaud et al. (2016) developed an Excel workbook that utilizes downloaded future climate scenario information from the online MarkSim weather file generator (http://gisweb.ciat.cgiar.org/MarkSimGCM/) to create CLIGEN input parameter files and WEPP model climate input files. This tool was used in Flanagan et al. (2014) to examine the projected future climate effects for a field with a silt loam soil in northeast Indiana under four types of land management systems. Flanagan et al. (2017) reported results of both changing temperatures and atmospheric CO_2 levels on WEPP model predicted crop yield, runoff, and soil loss for the mid- and late-twenty-first century.

8 Conclusion

Soil erosion by water processes have been identified and studied by many research scientists over the past 100 years. Physical processes include soil detachment by raindrops and shallow sheet flow, enhanced by raindrop impacts, and detachment by excess flow shear stress in rills and small channels. This chapter focused on rill, interrill, and ephemeral gully erosion processes, and the process-based WEPP model technology that can be used to estimate soil loss from these processes. A recent study indicated high model performance by WEPP when applied to a large set (1159 plot years) of validation data (Wang et al., 2023), even when run in uncalibrated mode (Nash-Sutcliffe model efficiencies: NSE = 0.82 for average annual runoff, and NSE = 0.72 for average annual soil loss). The WEPP model is a powerful tool that can be utilized to examine predicted runoff and soil loss under current baseline field conditions, and for some alternative conditions (different cropping, different tillage systems, other conservation practices, future climate inputs, etc.). Model development work is ongoing to improve various model components, and expand model capabilities.

9 Where to look for further information

Numerous resources are available that can provide more information on soil erosion by water processes research and erosion prediction technologies. A number of books have been published in the past 40 years related to the subject, including El-Swaify et al. (1985), Harmon and Doe (2001), Toy et al. (2002), Morgan (2004), Shukla (2011), and Lal (2017).

Within the United States, three national USDA laboratories have research programs that include soil erosion process measurement and modeling. These are:

- The National Soil Erosion Research Laboratory (NSERL), located on the Purdue University campus in West Lafayette, Indiana. https://www.ars.usda.gov/midwest-area/west-lafayette-in/national-soil-erosion-research/.
- The National Sedimentation Laboratory (NSL), located in Oxford, Mississippi. https://www.ars.usda.gov/southeast-area/oxford-ms/national-sedimentation-laboratory/.
- The National Laboratory for Agriculture and the Environment (NLAE), located on the Iowa State University campus in Ames, Iowa. https://www.ars.usda.gov/midwest-area/ames/nlae/.

International research locations with soil erosion research programs include the Institute of Soil and Water Conservation at the Northwest Agriculture and Forestry University in Yangling, China, the Katholieke Universiteit Leuven in Leuven, Belgium, Ghent University in Ghent, Belgium, Wageningen University in the Netherlands, the Commonwealth Scientific and Industrial Research Organisation (CSIRO) in Canberra, Australia, and Griffith University in Queensland, Australia.

The American Society of Agricultural and Biological Engineers (ASABE) sponsors a decennial international symposium on soil erosion research, with the most recent ones being in Puerto Rico (Flanagan and Thompson, 2023), Alaska (Flanagan et al., 2011), and Hawaii (Ascough and Flanagan, 2001). Proceedings for each of these with soil erosion research and modeling papers are available online. ASABE also publishes several journals that often include research findings on soil erosion and/or modeling (Journal of the ASABE - formerly Transactions of the ASABE, Applied Engineering in Agriculture, Journal of Natural Resources and Agricultural Ecosystems).

The Soil and Water Conservation Society (SWCS) in the United States includes soil erosion research scientists and soil conservation professionals, seeking to improve understanding and preservation of precious soil and water resources. Their publication, the *Journal of Soil and Water Conservation*, includes research findings on soil erosion processes and prediction technologies.

The World Association of Soil and Water Conservation (WASWAC) is a large international organization involved in activities that include soil erosion research and soil conservation practices. Their publication, the International Soil and Water Conservation Research (ISWCR) journal contains many articles related to soil erosion research.

The International Soil Conservation Organization (ISCO) holds international conferences every 2-3 years. Recent meetings have been held in New Delhi, India (2019), Lleida, Spain (2017), and El Paso, Texas (2015). ISCO information is available from the NSERL website at: https://topsoil.nserl.purdue.edu/isco/index.html. Current plans are for the next ISCO meeting to be held in 2026 at Purdue University and the NSERL in West Lafayette, Indiana. These conferences are excellent opportunities to meet with soil erosion research scientists and soil conservation practitioners from around the world.

10 References

Alberts, G. A., Nearing, M. A., Weltz, M. A., Risse, L. M., Pierson, F. B., Zhang, X. C., Laflen, J. M. and Simanton, J. R. (1995). Chapter 7. Soil component. In: Flanagan, D. C. and Nearing, M. A. (Eds). *USDA-Water Erosion Prediction Project: Hillslope Profile and Watershed Model Documentation*. NSERL Report No. 10, USDA-ARS National Soil Erosion Research Laboratory, West Lafayette, IN, pp. 7.1-7.47.

Al-Durrah, M. M. and Bradford, J. M. (1982). The mechanism of raindrop splash on soil surfaces. *Soil Sci. Soc. Am. J.* 46(5), 1086-1090.

Allen, R. G., Jensen, M. E., Wright, J. L. and Burman, R. D. (1989). Operational estimates of reference evapotranspiration. *Agron. J.* 81(4), 650-662.

Ascough, J. C. II and Flanagan, D. C. (Eds.) (2001). *Proceedings of the Soil Erosion Research for the 21st Century International Symposium*, 3-5 January 2001, Honolulu, Hawaii. American Society of Agricultural Engineers, St. Joseph, Michigan. 713 pp. https://elibrary.asabe.org/conference.asp?confid=se2001.

Ascough, J. C. II, Baffaut, C., Nearing, M. A. and Flanagan, D. C. (1995). Chapter 13. Watershed model channel hydrology and erosion processes. In: Flanagan, D. C. and Nearing, M. A. (Eds). *USDA-Water Erosion Prediction Project: Hillslope Profile and Watershed Model Documentation*. NSERL Report No. 10, National Soil Erosion Research Laboratory, USDA-Agricultural Research Service, West Lafayette, IN, pp. 13.1-13.20.

Ascough, J. C. II, Baffaut, C., Nearing, M. A. and Liu, B. Y. (1997). The WEPP watershed model: I. hydrology and erosion. *Trans. ASAE* 40, 921-933.

Carter, M. R. (2005). Conservation tillage. In: Hillel, D. (Ed). *Encyclopedia of Soils in the Environment*. Academic Press, Cambridge, MA, pp. 306-311.

Chu, S. T. (1978). Infiltration during an unsteady rain. *Water Resour. Res.* 14(3), 461-466.

Cruse, R., Flanagan, D., Frankenberger, J., Gelder, B., Herzmann, D., James, D., Krajewski, W., Kraszewski, M., Laflen, J. and Todey, D. (2006). Daily estimates of rainfall, water runoff, and soil erosion in Iowa. *J. Soil Water Conserv.* 61(4), 191-199.

David, O., Carlson, J. R., Yaege, L. A., Kipka, H., Flanagan, D. C. and Frankenberger, J. R. (2023). WEPP-GIS: Detailed assessment of soil erosion, runoff, and sedimentation in farm fields and small watersheds. In: Flanagan, D. C. and Thompson, A. M. (Eds). *Proceedings of the Soil Erosion Research under a Changing Climate International Symposium*. Abstract No. 23059. ASABE, St. Joseph, MI. https://doi.org/10.13031/soil.23059.

De Baets, S., Poesen, J., Meersman, J. and Serlet, L. (2011). Cover crops and their erosion-reducing effects during concentrated flow erosion. *CATENA* 85(3), 237-244.

de Roo, A. P. J., Wesseling, C. G. and Ritsema, C. J. (1996). LISEM: A single-event physically based hydrological and soil erosion model for drainage basins. I: Theory, input and output. *Hydrol. Proc.* 10(8), 1107-1117.

Dobre, M., Srivastava, A., Lew, R., Deval, C., Brooks, E. S., Elliot, W. J. and Robichaud, P. R. (2022). WEPPcloud: An online watershed-scale hydrologic modeling tool. Part II. Model performance assessment and applications to forest management and wildfires. *J. Hydrol.* 610, 127776. https://doi.org/10.1016/j.jhydrol.2022.127776.

Elliot, W. J. (2004). WEPP internet interfaces for forest erosion prediction. *J. Am. Wat. Resour. Assoc.* 40(2), 299-309.

Elliot, W. J. (2013). Erosion processes and prediction with WEPP technology in forests in the northwestern U.S. *Trans. ASABE* 56(2), 563-579.

Elliot, W. J. and Flanagan, D. C. (2023). Estimating WEPP cropland erodibility values from soil properties. *J. ASABE* 66(2), 329-351. https://doi.org/10.13031/ja.15218.

El-Swaify, S. A., Moldenhauer, W. C. and Lo, A. (Eds.) (1985). *Soil Erosion and Conservation*. Soil and Water Conservation Society, Ankeny, IA. 793 pp.

Finkner, S. C., Nearing, M. A., Foster, G. R. and Gilley, J. E. (1989). A simplified equation for modeling sediment transport capacity. *Trans. ASAE* 32, 1545-1550.

Flanagan, D. C. and Livingston, S. J. (Eds). (1995). *Water Erosion Prediction Project (WEPP) Version 95.7 User Summary*. NSERL Report No. 11, National Soil Erosion Research Laboratory, USDA-Agricultural Research Service, West Lafayette, IN, 139 pp.

Flanagan, D. C. and Nearing, M. A. (Eds). (1995). *USDA-Water Erosion Prediction Project (WEPP) Hillslope Profile a Watershed Model Documentation*. NSERL Report No. 10, National Soil Erosion Research Laboratory, USDA-Agricultural Research Service, West Lafayette, IN, 289 pp.

Flanagan, D. C. and Nearing, M. A. (2000). Sediment particle sorting on hillslope profiles in the WEPP model. *Trans. ASAE* 43(3), 573-583.

Flanagan, D. C. and Thompson, A. M. (Eds.) (2023). *Proceedings of the Soil Erosion Research Under a Changing Climate International Symposium*, 8-13 January 2023, Aguadilla, Puerto Rico, USA. American Society of Agricultural and Biological Engineers, St. Joseph, Michigan. https://elibrary.asabe.org/conference.asp?confid=soil2023.

Flanagan, D. C., Ascough, J. C. II and Nieber, J. L. (Eds.) (2011). *Proceedings of the International Symposium on Erosion and Landscape Evolution*, 18-21 September

2011, Anchorage, Alaska. American Society of Agricultural and Biological Engineers, St. Joseph, Michigan. https://elibrary.asabe.org/conference.asp?confid=isel2011.

Flanagan, D. C., Ascough, J. C. II, Nearing, M. A. and Laflen, J. M. (2001). Chapter 7. The Water Erosion Prediction Project (WEPP) model. In: Harmon, R. S. and Doe, W. W. III (Eds). *Landscape Erosion and Evolution Modeling*. Kluwer Academic Publishers/ Plenum Publishers, New York, NY, pp. 145–199.

Flanagan, D. C., Norton, L. D., Peterson, J. R. and Chaudhari, K. (2003). Using polyacrylamide to control erosion on agricultural and disturbed soils in rainfed areas. *J. Soil Water Conserv.* 58(5), 301-311.

Flanagan, D. C., Gilley, J. E. and Franti, T. G. (2007). Water Erosion Prediction Project (WEPP): Development history, model capabilities, and future enhancements. *Trans. ASABE* 50(5), 1603-1612.

Flanagan, D. C., Frankenberger, J. R., Cochrane, T. A., Renschler, C. S. and Elliot, W. J. (2013). Geospatial application of the Water Erosion Prediction Project (WEPP) model. *Trans. ASABE* 56(2), 591-601.

Flanagan, D. C., Trotochaud, J., Wallace, C. W. and Engel, B. A. (2014). Tool for obtaining projected future climate inputs for the WEPP and SWAT models. In: *Proceedings of the ASABE 21st Century Watershed Technology Conference*. November 1-7, 2014, University of Waikato, Hamilton, New Zealand. Paper No. 14-015. ASABE, St. Joseph, MI, 8 pp. https://doi.org/10.13031/wtcw.2014-015.

Flanagan, D. C., Frankenberger, J. R., Srivastava, A. and Trotochaud, J. (2017). Prediction technologies for assessment of climate change impacts. In: Simo, I., Poch, R. M. and Pla, I. (Eds). *Proceedings of the 1st World Conference on Soil and Water Conservation under Global Change - CONSOWA*. 12-16 June 2017, Lleida, pp. 391-394.

Flanagan, D. C., McGehee, R. P. and Srivastava, A. (2020). Evaluation of Different Precipitation Inputs to WEPP. Paper #2000740. ASABE, St. Joseph, MI, 18 pp. https://doi.org/10.13031/aim.202000740.

Flanagan, D. C., Frankenberger, J. R., Srivastava, A., Kucera, M. J. and Coreil, C. B. (2023). Water Erosion Prediction Project model for NRCS use. In: Flanagan, D. C., Thompson, A. M. (Eds). *Proceedings of the Soil Erosion Research under a Changing Climate International Symposium*, Paper No. 23002. ASABE, St. Joseph, MI, 8 pp. https://doi.org/10.13031/soil.23002.

Foote, G. B. and du Toit, P. S. (1969). Terminal velocity of raindrops aloft. *J. Appl. Meteorol.* 8(2), 249–253.

Foster, G. R. and Meyer, L. D. (1975). Mathematical simulation of upland erosion by fundamental erosion mechanics. In: Present and prospective technology for predicting sediment yields and sources. ARS-S-40. USDA-ARS, US Gov. Print. Office, pp. 190–207.

Foster, G. R., Lane, L. J., Nowlin, J. D., Laflen, J. M. and Young, R. A. (1980). Chapter 3. A model to estimate sediment yield from field-size areas: Development of model. In: Knisel, W. G. (Ed). *CREAMS: A Field-Scale Model for Chemicals, Runoff, and Erosion from Agricultural Management Systems*. Conservation Res. Report No. 26. USDA Science and Education Administration, Washington, DC, pp. 36–64.

Foster, G. R., Flanagan, D. C., Nearing, M. A., Lane, L. J., Risse, L. M. and Finkner, S. C. (1995). Chapter 11. Hillslope erosion component. In: Flanagan, D. C. and Nearing, M. A. (Eds). *USDA-Water Erosion Prediction Project: Hillslope Profile and Watershed*

Model Documentation. NSERL Report No. 10, USDA-ARS National Soil Erosion Research Laboratory, West Lafayette, IN. pp. 11.1-11.12.

Fox, D. M. and Bryan, R. B. (2000). The relationship of soil loss by interrill erosion to slope gradient. *CATENA* 38(3), 211-222.

Gabbard, D. S., Huang, C., Norton, L. D. and Steinhardt, G. C. (1998). Landscape position, surface hydraulic gradients and erosion processes. *Earth Surf. Proc. Land.* 23(1), 83-93.

Gelder, B., Skelnar, T., James, D., Herzmann, D., Cruse, R., Gesch, K. and Laflen, J. (2018). The Daily Erosion Project - Daily estimates of water runoff, soil detachment, and erosion. *Earth Surf. Proc. Land.* 43(5), 1105-1117.

Gilley, J. E., Kottwitz, E. R. and Simanton, J. R. (1990). Hydraulic characteristics of rills. *Trans. ASAE* 33, 1900-1906.

Gilley, J. E. and Weltz, M. A. (1995). Chapter 10. Hydraulics of overland flow. In: Flanagan, D. C. and Nearing, M. A. (Eds). *USDA-Water Erosion Prediction Project: Hillslope Profile and Watershed Model Documentation*. NSERL Report No. 10, USDA-ARS National Soil Erosion Research Laboratory, West Lafayette, IN, pp. 10.1-10.7.

Guy, B. T., Dickinson, W. T. and Rudra, R. P. (1987). The roles of rainfall and runoff in the sediment transport capacity of interrill flow. *Trans. ASAE* 30, 1378-1386.

Harmon, R. S. and Doe, W. W. II (Eds.) (2001). *Landscape Erosion and Evolution Modeling*. Kluwer Academic / Plenum Publishers, New York. 540 pp.

Huang, C.-H. and Laflen, J. M. (1996). Seepage and soil erosion for a clay loam soil. *Soil Sci. Soc. Am. J.* 60(2), 408-416.

Kar, S. K., Kumar, S., Sankar, M., Patra, S., Singh, R. M., Shrimali, S. S. and Ojasvi, P. R. (2022). Process-based modelling of soil erosion: Scope and limitation in the Indian context. *Curr. Sci.* 122(5), 533-541.

Kinnell, P. I. A. (1981). Rainfall intensity-kinetic energy relationships for soil loss prediction. *Soil Sci. Soc. Am. J.* 45(1), 153-155.

Kinnell, P. I. A. (1991). The effect of flow depth on sediment transport induced by raindrops impacting shallow flows. *Trans. ASABE* 34(1), 161-168.

Kinnell, P. I. A. (2022). Insights into the sediment transport processes operating in rain-impacted flows. *CATENA* 217, 106452.

Knisel, W. G. (Ed). (1980). *CREAMS: A Field-Scale Model for Chemicals, Runoff, and Erosion from Agricultural Management Systems*. Conservation Res. Report No. 26. USDA Science and Education Administration, Washington, DC.

Knisel, W. G. and Douglas-Mankin, K. R. (2012). CREAMS/GLEAMS: Model use, calibration, and validation. *Trans. ASABE* 55(4), 1291-1302.

Kucera, M. J. and Coreil, C. B. (2023). Overview of USDA-NRCS erosion prediction technology and Conservation Resources Land Management Operations Database and use. In: Flanagan, D. C. and Thompson, A. M. (Eds). *Proceedings of the Soil Erosion Research under a Changing Climate International Symposium*, Paper No. 23007. ASABE, St. Joseph, MI, 8 pp. https://doi.org/10.13031/soil.23007.

Lal, R. (Ed.) (2017). *Encyclopedia of Soil Science* (3rd edn.). CRC Press, Boca Raton, FL. 2653 pp.

Lew, R., Dobre, M., Srivastava, A., Brooks, E. S., Elliot, W. J., Robichaud, P. R. and Flanagan, D. C. (2022). WEPPcloud: An online watershed-scale hydrologic modeling tool. Part I. Model description. *J. Hydrol.* 608, 127603. https://doi.org/10.1016/j.jhydrol.2022.127603.

Lindley, M. R., Barfield, B. J. and Wilson, B. N. (1995). Chapter 14. Surface impoundment element model description. In: Flanagan, D. C. and Nearing, M. A. (Eds). *USDA-Water Erosion Prediction Project: Hillslope Profile and Watershed Model Documentation*, NSERL Report No. 10, National Soil Erosion Research Laboratory, USDA-Agricultural Research Service, West Lafayette, IN, pp. 14.1–14.28.

Lindley, M. R., Barfield, B. J., Ascough, J. C. II, Wilson, B. N. and Stevens, E. W. (1998a). Hydraulic simulation techniques incorporated in the surface impoundment of WEPP. *App. Eng. Agric.* 14, 249–256.

Lindley, M. R., Barfield, B. J., Ascough, J. C. II, Wilson, B. N. and Stevens, E. W. (1998b). The surface impoundment element for WEPP. *Trans. ASAE* 41(3), 555–564.

McGehee, R. P., Flanagan, D. C. and Srivastava, P. (2020). WEPPCLIFF: A command-line tool to process climate inputs for soil loss models. *J. Open Source Softw.* 5(49), 2029. https://doi.org/10.21105/joss.02029.

McGehee, R. P., Flanagan, D. C., Srivastava, P. and Nearing, M. A. (2021). Chapter 16. Rainfall erosivity: Essential historical, conceptual, and practical perspectives for continued application. In: Comino, J. R. (Ed). *Precipitation*. Elsevier, Amsterdam, pp. 373–394. https://doi.org/10.1016/B978-0-12-822699-5.00014-8.

McGehee, R. P., Flanagan, D. C., Srivastava, P., Engel, B. A., Huang, C. and Nearing, M. A. (2022). An updated isoerodent map of the conterminous United States. *Intl. Soil Water Conserv. Res.* 10(1), 1–16.

McGehee, R. P., Flanagan, D. C. and Engel, B. A. (2023). A WEPP-water quality model for simulating nonpoint source pollutants in nonuniform agricultural hillslopes: Model development and sensitivity. *Intl. Soil Water Conserv. Res.* 11(3), 455–469. https://doi.org/10.1016/j.iswcr.2023.02.002.

Mein, R. G. and Larson, C. L. (1973). Modeling infiltration during a steady rain. *Water Resour. Res.* 9(2), 384–394.

Meyer, L. D., Foster, G. R. and Romkens, M. J. M. (1975). Source of soil eroded by water from upland slopes. In: *Present and Prospective Technology for Predicting Sediment Yields and Sources*. ARS-S-40. USDA-ARS, US Gov. Print. Office, pp. 177–189.

Meyer, L. D. (1981). How rain intensity affects interrill erosion. *Trans. ASAE* 24, 1472–1475.

Monteith, J. L. (1965). Evaporation and environment. *Symp. Soc. Exp. Biol.* 19, 205–234.

Morgan, R. P. C. (2004). *Soil Erosion and Conservation* (3rd edn.). Wiley-Blackwell, Hoboken, NJ. 320 pp.

Morgan, R. P. C., Quinton, J. N., Smith, R. E., Govers, G., Poesen, J. W. A., Auerswald, K., Chisci, G., Torri, D. and Styczen, M. E. (1998). The European Soil Erosion Model (EUROSEM): A dynamic approach for predicting sediment transport from fields and small catchments. *Earth Surf. Proc. Land.* 23(6), 527–544.

Nearing, M. A., Foster, G. R., Lane, L. J. and Finkner, S. C. (1989). A process-based soil erosion model for USDA-Water Erosion Prediction Project technology. *Trans. ASAE* 32(5), 1567–1593.

Nearing, M. A., Norton, L. D., Bulgakov, D. A., Larionov, G. A., West, L. T. and Dontsova, K. M. (1997). Hydraulics and erosion in eroding rills. *Water Resour. Res.* 33(4), 865–876.

Nicks, A. D., Lane, L. J. and Gander, G. A. (1995). Chapter 2. Weather generator. In: Flanagan, D. C. and Nearing, M. A. (Eds). *USDA-Water Erosion Prediction Project: Hillslope Profile and Watershed Model Documentation*, NSERL Report No. 10, National Soil Erosion Research Laboratory, USDA-Agricultural Research Service, West Lafayette, IN, pp. 2.1–2.22.

Onstad, C. A. (1984). Depression storage on tilled soil surfaces. *Trans. ASAE* 27, 729–732.

Renschler, C. S. (2003). Designing geo-spatial interfaces to scale process models: The GeoWEPP approach. *Hydrol. Proc.* 17(5), 1005-1017.

Renschler, C. S. and Zhang, H. (2020). Long-term, process-based, continuous simulations for a cluster of six smaller, nested rangeland watersheds near Tombstone, AZ (USA): Establishing a baseline for event-based runoff and sediment yields. *Sci. Total Environ.* 717, 137089. https://doi.org/10.1016/j.scitotenv.2020.137089.

Revuelta-Acosta, J. D., Flanagan, D. C., Engel, B. A. and King, K. W. (2021). Improvement of the Water Erosion Prediction Project (WEPP) model for quantifying subsurface drainage flow at field scale. *Agric. Water Manag.* 244, 106597. https://doi.org/10.1016/j.agwat.2020.106597.

Rose, C. W., Williams, J. R., Sander, G. C. and Barry, D. A. (1983). A mathematical model of soil erosion and deposition processes: I. Theory for a plane land element. *Soil Sci. Soc. Am. J.* 47(5), 991–995.

Rose, C. W., Coughlan, K. J., Ciesiolka, C. A. A. and Fentie, B. (1997). Program GUEST (Griffith University Erosion System Template). In: Coughlan, K. J. and Rose, C. W. (Eds). *A New Soil Conservation Methodology and Application to Cropping Systems in Tropical Steep Lands*. ACIAR Tech. Report, No. 40, Canberra. Australian Centre for International Agricultural Research, Canberra, pp. 34–58.

Savabi, M. R., Flanagan, D. C., Frankenberger, J. R., Hubbard, R. K., Bosch, D. D. and Potter, T. L. (2011). Development of a WEPP-Water Quality (WEPP-WQ) model. In: Flanagan, D. C., Ascough, J. C. II and Nieber, J. A. (Eds). *CD-ROM Proceedings International Symposium on Erosion and Landscape Evolution (ISELE)*, Paper No. 11146. ASABE, St. Joseph, MI.

Shukla, M. K. (Ed.) (2011). *Soil Hydrology, Land Use and Agriculture*. CAB International, Wallingford, Oxfordshire, UK. 434 pp.

Srivastava, A., Flanagan, D. C., Frankenberger, J. R. and Engel, B. A. (2019). Updated climate database and impacts on WEPP model predictions. *J. Soil Water Conserv.* 74(4), 334–349.

Srivastava, A., McGehee, R. P., Flanagan, D. C., Frankenberger, J. R. and Engel, B. A. (2023). Modeling the impacts of climate change on crop yield, runoff, and soil loss using the Water Erosion Prediction Project (WEPP) model. In: Flanagan, D. C. and Thompson, A. M. (Eds). *Proceedings of the Soil Erosion Research under a Changing Climate International Symposium*. Paper No. 23075. ASABE, St. Joseph, MI, 7 pp. https://doi.org/10.13031/soil.23075.

Stockle, C. O., Williams, J. R., Rosenberg, N. J. and Jones, C. A. (1992). A method for estimating the direct and climatic effects of rising atmospheric carbon dioxide on growth and yield of crops: Part I-Modification of the EPIC model for climate change analysis. *Agric. Syst.* 38(3), 225–238.

Stone, J. J., Lane, L. J. and Shirley, E. D. (1992). Infiltration and runoff simulation on a plane. *Trans. ASAE* 35, 161–170.

Stone, J. J., Lane, L. J., Shirley, E. D. and Hernandez, M. (1995). Chapter 4. Hillslope surface hydrology. In: Flanagan, D. C. and Nearing, M. A. (Eds). *USDA-Water Erosion Prediction Project: Hillslope Profile and Watershed Model Documentation*. NSERL Report No. 10, National Soil Erosion Research Laboratory, USDA-Agricultural Research Service, West Lafayette, IN, pp. 4.1–4.20.

Toy, T. J., Foster, G. R. and Renard, K. G. (2002). *Soil Erosion: Processes, Prediction, Measurement, and Control*. John Wiley and Sons, Inc., New York. 352 pp.

Trotochaud, J., Flanagan, D. C. and Engel, B. A. (2016). A simple technique for obtaining future climate data inputs for natural resource models. *Appl. Eng. Agric.* 32(3), 371-381.

Wang, L., Flanagan, D. C. and Cherkauer, K. A. (2017). Development of a coupled water quality model. *Trans. ASABE* 60(4), 1153-1170.

Wang, S., Flanagan, D. C., Engel, B. A. and Zhou, N. (2020). Impacts of subsurface hydrologic conditions on rill sediment transport capacity. *J. Hydrol.* 591, 125582.

Wang, S., Flanagan, D. C., Engel, B. A. and McIntosh, M. M. (2021). Impacts of subsurface hydrologic conditions on sediment selectivity and sediment transport in rills. *CATENA* 207, 105703.

Wang, S., McGehee, R. P., Guo, T., Flanagan, D. C. and Engel, B. A. (2023). Calibration, validation, and evaluation of the Water Erosion Prediction Project (WEPP) model for hillslopes with natural runoff plot data. *Intl. Soil Water Conserv. Res.* 11(4), 669-687. https://doi.org/10.1016/j.iswcr.2022.10.004.

Wischmeier, W. H. and Smith, D. D. (1978). *Predicting Rainfall Erosion Losses - A Guide to Conservation Planning*. Agriculture Handbook No. 537, USDA - Science and Education Administration. U.S. Government Printing Office, Washington, DC, 58 pp.

Yalin, Y. S. (1963). An expression for bedload transportation. *J. Hydraul. Div. Proc. ASCE* 89(HY3), 221-250.

Young, R. A. and Wiersma, J. L. (1973). The role of raindrop impact in soil detachment and transport. *Water Resour. Res.* 9(6), 1629-1636.

Zhang, H., Renschler, C. S., Nichols, M. H. and Nearing, M. A. (2021). Long-term, process-based, continuous simulations for a small, nested rangeland watershed near Tombstone, AZ (USA): Extending model validity to include soil redistribution. *Sci. Total Environ.* 792, 148403. https://doi.org/10.1016/j.scitotenv.2021.148403.

Zhang, X. C. and Wang, Z. L. (2017). Interrill soil erosion processes on steep slopes. *J. Hydrol.* 548, 652-664.

Zhang, X. C., Zheng, F. L., Chen, J. and Garbrecht, J. D. (2020). Characterizing detachment and transport processes of interrill soil erosion. *Geoderma* 376, 114549.

Zhang, X. J. (2023). Role of raindrop impact in detachment and transport processes of interrill soil erosion. *Intl. Soil Water Conserv. Res.* 11(4), 592-601. https://doi.org/10.1016/j.iswcr.2022.11.001.

Chapter 3

Assessing the impact of climate change on soil erosion by water

Karl Auerswald, Technical University of Munich, Germany; and Peter Fiener, University of Augsburg, Germany

1 Introduction

Climate change exerts a significant impact on the erosive potential of rainfall. This potential is commonly referred to as rain erosivity (Fig. 1). However, climate change also alters the counterbalancing ability of crops to mitigate soil erosion by providing a protective soil cover. The effectiveness of this protective cover relies on the alignment or discrepancy between the seasonal distribution of rain erosivity and the seasonal distribution of soil cover. Climate change influences both the timing of erosive rainfall events and the development of vegetation, including crops, and may also alter agricultural management (type of crops and crop rotations, soil management, etc.).

The impact of climate change varies in intensity from region to region. While temperatures generally rise, this increase is notably more significant in polar than in tropical regions, which is called artic amplification (Rantanen et al., 2022). Even more diverse are the effects on precipitation. It is assumed that a one-degree increase in temperature will result in a relatively 2–3% increase in evapotranspiration and subsequently in precipitation on a global average (Roderick et al., 2014). However, certain areas may experience pronounced precipitation changes in either direction due to altering atmospheric circulation patterns (e.g. Odemark et al., 2023). Additionally, in many regions, snowfall may (partly) be replaced by rainfall and snowmelt occurs earlier (Crossman et al.,

http://dx.doi.org/10.19103/AS.2023.0131.05

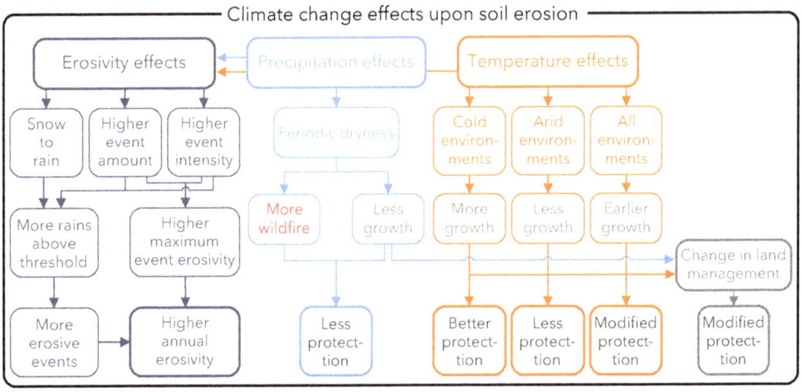

Figure 1 Overview of the many impacts of climate change on soil erosion by water.

2013). As a result, it is apparent that there is no single manifestation of climate change but rather several. Moreover, the impacts of temperature increase will vary in dry or wet areas and in cold or warm regions (Fig. 1). They may lead to more growth in cold areas (Wang et al., 2023) and less growth in arid areas, where shifts can be particularly catastrophic (Scheffer et al., 2001). Hence the protection of soil by plants may be improved or impaired.

In addition to the general trends of rising temperature and increasing precipitation, a major consequence of climate change is that certain weather conditions, either wet or dry, are prolonged due to the increased amplitude of Rossby waves (Coumou et al., 2015). This results in increased variation of precipitation patterns in most regions (Dore, 2005). Longer wet periods cause more runoff and more erosion. The loss of water due to runoff may in turn cause water shortage, in particular, because longer dry periods increase the probability of periodic droughts. Droughts will restrict plant growth and decrease the soil protection by plants. Protection by plants is particularly reduced when wildfires become more frequent (Westerling et al., 2006; Pausas and Fernández-Muñoz, 2012). On top of these multiple influences, potential droughts and changing phenological rhythms of crops (Menzel et al., 2020) will impact farmers' options and decisions that in turn will affect erosion. For instance, irrigation to alleviate drought can cause additional erosion. Some of the longest rill erosion lengths described in the scientific literature were found under center-pivot irrigation systems (Batista et al., 2023).

Given this complex situation, we will focus on the impacts that are likely to occur in significant portions of the temperate zone where arable land prone to erosion is prevalent while acknowledging that we are excluding large parts of the globe (the tropics, the polar and subpolar regions). Additionally, we recognize that even within this well-defined zone, peculiarities may exist that could result in different behavior than what is described in this chapter.

The increased variability in rainfall has to be counterbalanced via an appropriate ecosystem water storage to avoid potentially negative effects. Here soils are of tremendous importance due to their ability to serve as buffers, typically assessed as plant-available water capacity within the rooting zone. This functionality becomes impaired as a consequence of erosion truncating the rooting zone (Pierce et al., 1983; Lobo et al., 2005). On a landscape scale, the loss in functionality can only partly be compensated through deeper soils in depositional areas (Quinton et al., 2022). Soil erosion, as we will demonstrate, is commonly exacerbated by climate change. Consequently, a conflict arises from the fact that climate change increasingly undermines the functionality of soils while simultaneously intensifying the demand for their proper functioning.

2 Influence of climate change on event and annual erosivity

2.1 Fundamental background

The erosive potential of rainfall is typically assessed using the *R* factor proposed by Wischmeier (1959) and Wischmeier and Smith (1958) for the Universal Soil Loss Equation (USLE). Event erosivity (also called erosion index) is determined by the cumulative kinetic energy of a rainfall event and its largest intensity within a 30-min interval (Equation 1). Subsequently, the long-term annual erosivity, called *R* factor, is derived from the sum of event erosivities over many years divided by the number of years.

$$EI = E_{kin} \times I_{max,\,30} \qquad 1$$

Variables (all per unit area and rain event) are: EI event erosivity or erosion index, E_{kin} kinetic energy, $I_{max,\,30}$ maximum 30-min intensity of a rainfall event.

Not all rain events cause erosion. According to Wischmeier and Smith (1978), rains of less than 12.4 mm are regarded as non-erosive unless they deliver more than 6.4 mm within 15 min. Slightly modified thresholds were used in other studies (e.g. Wischmeier, 1959; Rogler and Schwertmann, 1981), but generally, these thresholds reflect that it requires relatively heavy rainfall to cause surface runoff and soil erosion. This is important in relation to climate change because erosivity can change even if the amount of rain remains unchanged. If, for example, the rainfall amount of two smaller events falls during one event, a considerably higher erosivity will result. Monthly, seasonal, or annual amounts of rain are thus unsuitable to detect changes in rain erosivity if climate changes the amount and number of individual events but not the sum over the respective period.

The kinetic energy of rain is dependent on the raindrop size distribution, the mass of the different drop sizes, and their associated falling velocity (v). The total rain's mass (m) can be determined by multiplying rain intensity by rain duration. As a result, the intensity is squared due to the multiplication of $E_{kin}\left(=\frac{1}{2}\times m\times v^2\right)$ and $I_{max,30}$ during the calculation of the erosion index. Additionally, intense rains have larger drops (Laws and Parsons, 1943) that fall at a greater velocity (Jones et al., 2010) giving even more weight to intensity. Thus, the erosion index and the R factor are highly sensitive to changes in rainfall intensity.

2.2 Past changes of the R factor

Due to the unaltered energy input from radiation, climate change has only a small impact on globally averaged evaporation and, consequently, on precipitation. It is anticipated that both evaporation and precipitation will rise by a mere 2–3% per kelvin in response to temperature rise (Roderick et al., 2014; Bürger et al., 2014). However, the effects on individual rainfall events are significantly more pronounced. Climate change affects four crucial properties of (convective) rainfall events, which in consequence affect (annual) erosivity:

1. Increasing rainfall amount per event rain: The saturation vapor pressure rises by approximately 7% per kelvin, which is called CC rate because this prediction results from the Clausius–Clapeyron equation (Roderick et al., 2014; Bürger et al., 2014). While annual rain increases by only 2–3% per kelvin, the increase for individual rains should be 7% per kelvin. However, empirical observations reveal that during intense precipitation events, the increase can reach 14% within time intervals of hours, which is referred to as the super-CC rate (Westra et al., 2014). This effect is particularly pronounced at temperatures ranging from 12°C to 22°C (Lenderink and van Meijgaard, 2008). The intensification of rainfall can be attributed to an intensification of cloud dynamics (Loriaux et al., 2013; Westra et al., 2014) sometimes coupled with the transition from stratiform to convective regimes (Haerter and Berg, 2009). This intensification arises from higher surface temperatures and, in particular, from the higher release of sensible heat during condensation of moister air. This additional energy augments the updraft strength of the convective cell (Loriaux et al., 2013; Westra et al., 2014), leading to an increased influx of moist air into the convection cell and, thus, surpassing the CC rate in terms of precipitation amount.
2. Amplification of rainfall intensity and drop size: The increase in precipitation amount also amplifies intensity when the cell's lifespan remains unchanged. Moreover, higher surface temperatures and increased energy release during condensation elevate the convective

updraft through a rise in the environmental lapse rate and a decrease in the saturated adiabatic lapse rate (Bürger et al., 2014). As a result, rainfall forms more rapidly, causing the intensity to increase even further in comparison to the precipitation amount. Simultaneously, the intensified upward air currents impede the falling of small raindrops, thereby fostering the additional enlargement of raindrop size (Bürger et al., 2014) and kinetic energy. Since both precipitation amount and intensity are multiplied during the calculation of the erosion index, these combined effects lead to a superproportional augmentation of the erosion index.

3. Increased frequency of erosive events: This phenomenon can have two different origins. (i) With increasing temperatures, the potential for thunderstorms and convective rainfall intensifies. Due to the rise in precipitation amount and intensity, more rainfall events surpass the thresholds for erosive rainfall. The findings regarding a heightened frequency of heavy rain, however, are variable and point to a large influence of the methods and definitions applied during the studies (Mohr and Kunz, 2013; Mueller and Pfister, 2011; Deumlich and Gericke, 2020). (ii) With higher temperature, snowfall may be replaced by rainfall. This rainfall may be erosive, while erosion by snowmelt may decrease. Both options are possible but do not necessarily occur in combination because both forms of erosion are governed by different principles. Rainfall erosion depends on how much rain falls during an event, while snowmelt erosion depends on how much snow has accumulated during a frost period and how fast it is melted after the frost period.
4. Diminished wind velocities: The projected climate change scenarios indicate a likelihood of longer periods of reduced wind velocities due to altered weather patterns (Coumou et al., 2014; Mann et al., 2018). The global wind systems are fueled by the temperature gradient between the tropics and the polar regions, which is expected to diminish due to climate change. This reduction is attributed to the faster and greater warming of the polar regions compared to the tropics (Rantanen et al., 2022), caused by ice retreat and the coverage of ice with melting water, leading to reduced albedo and increased radiation absorption (Greuell, 2000) and other mechanisms. Moreover, the weakening of summer circulation, which decreases eddy kinetic energy and increases the amplitude of Rossby waves, prolongs specific weather conditions (Coumou et al., 2015). Extended periods of rainfall will contribute to a greater number of events surpassing the thresholds of the *R* factor and becoming erosive. Furthermore, convective rain cells typically exhibit movement, limiting the amount of rainfall delivered to a specific location. A decrease in wind velocity consequently leads to an increase in rainfall

amount at a location crossed by a convective cell and, subsequently, an elevation in *R*, even if the amount of rain released by a given cell remains unchanged.

All four aforementioned changes act multiplicatively on the *R* factor. Even slight increases in air temperature can thus have strong effects on rain erosivity. It is thus not surprising that erosivity has about doubled in Germany during the past 60 years (Fiener et al., 2013; Auerswald et al., 2019b) and closely follows the trend of global temperature (Auerswald et al., 2019b; Fig. 2, left panel).

2.3 Future changes – R factor of climate projections

Climate projections derived from climate models currently provide only daily precipitation and daily temperatures, which poses a limitation as it renders the application of the erosion index equation unfeasible. The use of proxies, like the Modified Fournier Index (Arnoldus, 1977; Chen et al., 2023) or the rater similar Precipitation Concentration Index (Oliver, 1980) or other somewhat more sophisticated approaches (Diodato et al., 2017) that rely on the ratios of long-term mean monthly and annual precipitation, cannot be effective. Already Arnoldus (1977) had noted that the correlation equation differs from region to region and has to be calibrated for the specific conditions where it is used. If monthly,

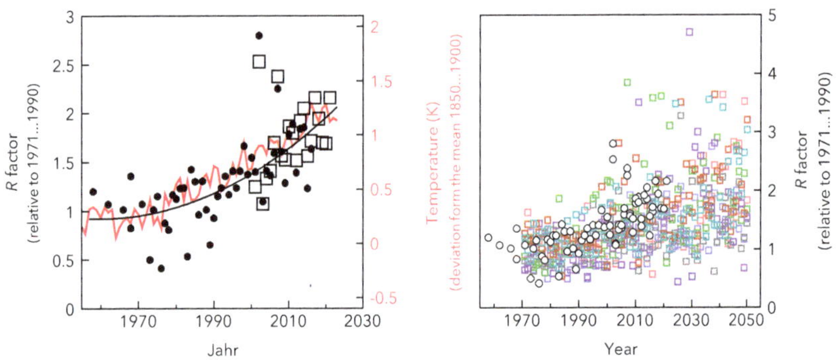

Figure 2 Change of erosivity in Germany over time relative to erosivity from 1971 to 1990 Left panel: measured erosivities and comparison with the global change in temperature (black circles result from high-resolution measurements at 50–115 meteorological stations); squares result from radar measurement with 5-min resolution for ca. 360 000 raster cells of 1 km² size; the erosivity data were taken from Auerswald et al. (2019b) before 2018 and from Winterrath (2022) for 2018 to 2021; the black line is a quadratic regression (R^2 = 0.48; P < 0.0001); the temperature data (red line) were taken from Met Office Hedlay Center (2023). Right panel: *R* factors derived from climate models for the representative concentration pathway RCP 8.5, which is a business-as-usual scenario (color markers; each color represents one of seven approved models) and comparison with the measured data from the left panel (white circles).

seasonal, and annual precipitation increase as projected by 2–3% per kelvin, such approaches hardly register any change. This would lead to the expectation of minor changes in the *R* factor. However, this perspective fails to account for the fact, as demonstrated earlier, that alterations in event characteristics are much more significant than changes in average precipitation. Consequently, any calibration function required to convert an average rainfall amount index into *R* must change over time. A notable example of this can be observed in Bavaria. Rogler and Schwertmann (1981) published *R* factors for 18 meteorological stations based on measurements conducted mainly in the 1960s and 1970s. Fifteen of them were still operational after 2000. Utilizing these *R* factors by Rogler and Schwertmann (1981) and the corresponding Modified Fournier Index from that period yielded a very close calibration function (r^2 = 0.92; Fig. 3, left panel). However, when deriving the *R* factors and the Modified Fournier Indices of the same stations for the years 2001 to 2017, a narrow calibration function was again obtained (r^2 = 0.92; Fig. 3, left panel) but with a slope almost twice as steep. Employing the old calibration function would have resulted in an underestimation of *R* for the period 2001 to 2017 by a factor of almost two. Note that the Modified Fournier Index between both periods hardly changed because of the minor changes in precipitation, while *R* changed considerably. Given that it is impossible to

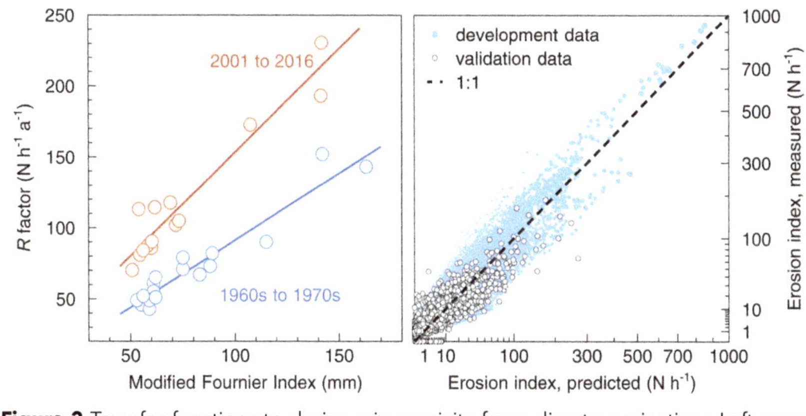

Figure 3 Transfer functions to derive rain erosivity from climate projections Left panel: Correlation of the Modified Fournier Index with the long-term mean annual *R* factor of 18 meteorological stations with data obtained in the 1960s and 1970s (past) and data obtained from 2001 to 2016 (present). Past erosivities were taken from Rogler and Schwertmann (1981). Present data were obtained at the same meteorological stations (1 min temporal resolution) except for three of the 18 original stations that were not operational anymore. Right panel: Correlation of the measured erosion index with a prediction using the measured daily amount of rain and the daily air temperature (605 738 erosive events were used for developing the transfer function and 15 133 erosive events were used for validation; the event erosivity was calculated from rain data with 5-min resolution; adopted from Auerswald et al., 2019c). The y-axis is square root scaled. The unit N h^{-1} can be converted to MJ mm ha^{-1} h^{-1} by multiplication by 10.

execute calibrations for projections, employing the Modified Fournier Index or similar indices for climate projections fails. It is thus not surprising that Biasutti and Seager (2015) claim that progress could be made by establishing a more reliable functional relationship between daily rainfall and erosivity.

Considering the substantial influence of air temperature on precipitation (Lenderink and van Meijgaard, 2008) and event erosivity, as demonstrated earlier, a more promising approach is to establish correlations between event erosivity and daily temperature and daily precipitation instead of long-term sums. Indeed, such an approach yielded a strong correlation (Fig. 3, right panel) for the following transfer function using daily precipitation (P_d in mm) and daily mean air temperature (T_a in °C) for temperatures above 0°C:

$$EI = 1.57 \times 10^{-6} \times P_d^{2.63} \times T_a^{2.52} \quad (2)$$

This equation can be applied to climate projections that provide daily temperature and precipitation data. It has to be noted, however, that climate projections have more difficulties in estimating future precipitation than future temperature (Slingo et al., 2022).

Auerswald et al. (2019c) have employed this methodology for the hydrological catchment area of south Germany, which encompasses significant portions of Austria, Switzerland, Italy, and other neighboring countries. They utilized seven climate models (see Appendix) approved by the German Weather Service (Hübener et al., 2017), each with a spatial resolution of 0.11° (approximately 12 km resulting from different CORDEX regional climate models (RCM) for the European domain (EURO-CORDEX, Jones et al., 2011; Knist et al., 2017). All projections (see Appendix) used the business-as-usual scenario following a representative concentration pathway (RCP) leading to a radiative forcing of 8.5 W m^{-2} by 2100 (RCP8.5), corresponding to 1370 ppm CO_2 equivalent (for a detailed overview of the development of the RCP scenarios see van Vuuren et al., 2011). When comparing R estimated from the climate projections with the measured trend of R for the period 1970 to 2022, a close alignment was observed (Fig. 2, right panel), even though both datasets were determined independently and with different equations, namely Equation 1 and Equation 2. Projected R values indicate that erosivity will likely continue to increase in the future. Moreover, Fig. 2 shows increasing fluctuations between years, allowing for events that surpass existing experience. The Ahrtal flood in 2021 in Germany, which resulted in at least 180 fatalities (Mohr et al., 2023), is a foretaste of what could potentially occur in the future.

2.4 Outlook on convection-permitting climate simulations

For assessing the future development of climate change, we rely on climate projections. Models used for climate projections have three pronounced limitations regarding their usefulness for deriving rain erosivity: (1) Presently, projections do not explicitly consider convective rainfall, which is the major cause of erosion events. (2) They produce only daily rainfall while sub-daily resolution is needed to calculate erosivity, preferably even sub-hourly resolution. (3) Their spatial resolution even after downscaling is too low (often in the range of 10 km) while we know that rain erosivity can change enormously within only a few hundred meters (Fiener and Auerswald, 2009; Fischer et al., 2016). This is due to the steep spatial gradients in rain intensity in convective rainstorms (Lochbihler et al., 2017). Convective rain cells hardly exceed a width of 1 km, which implies that steep gradients must exist within 1 km. These steep gradients are only somewhat attenuated by the movement of the convective rain cells, which increases their spatial coverage (for an illustration, see Fischer et al., 2016). The attenuations of intensity caused by low temporal resolution and by a low spatial resolution and a possible attenuation caused by the simulation or measuring methods lead to an underestimation of erosivity and need to be corrected (Fischer et al., 2018).

Only recently, the development of convection-permitting climate simulations started to overcome the first and partly the other two limitations by providing rainfall at hourly resolution and also in a finer but still insufficient spatial resolution. Thus, convection-permitting climate simulations offer an enormous potential for assessing the erosivity of a future climate without requiring to derive erosivity from daily rainfall as in the approach illustrated in the previous chapter. The improvements in convection-permitting climate simulations result from a better representation of the diurnal cycle (Prein et al., 2015) and of the return periods (Rybka et al., 2022). However, these simulations are time-consuming and require a considerable amount of computing power, which limits their application to small areas and short intervals of time.

Chapman et al. (2021) used convection-permitting climate simulations for the first time to calculate rainfall erosivity in Tanzania and Malawi for eight years in the past and eight years in the future. The spatial and temporal cover of COSMO-CLM (Rockel et al., 2008) is considerably broader covering Central Europe and 109 years. Uber et al. (2023) used rainfall from COSMO-CLM and the temporal scaling factor developed by Fischer et al. (2018) to correct for the hourly resolution but did not correct for the spatial resolution (3 km), which was also not sufficiently resolved to prevent a loss in peak intensities. This way, Uber et al. (2023) could show that rainfall erosivity in Central Europe is likely to increase by 84% when comparing 2071-2100 with 1971-2000. The derived increase from 1977-2000 to 2001-2019 was still considerably lower than what has been measured (Fig. 2) for this time interval, but the use of convection-permitting simulations is still a major improvement in deriving rain erosivity from climate projections. Developments of convection-permitting climate simulations like IKON-CLM (Pham et al., 2021) may further improve our estimates of future developments in rain erosivity.

3 Influence of climate change on the protective effect of crops

3.1 Fundamental background

The impact of crops stems from their ability to shield the soil surface from raindrop impact. Fewer soil particles become detached and transportable, and the raindrop-induced particle transport during sheet and rill erosion (Kinnell, 1991) is impeded. Most of the effect of plants to protect soils against erosion can be attributed to the presence of soil cover, although other influences, such as rooting density and microbial activity, also contribute to the effect of plants (for a quantification of the different effects, see Yoder et al., 1997). In the interest of brevity, our focus will be on plant cover, as it is anticipated that the most substantial effects induced by climate change will be associated with changes in cover.

The occurrence of soil erosion on a bare surface is contingent upon the presence or absence of erosive rainfall. Consequently, the protection of a soil surface by a crop relies upon both the seasonal variation in rain erosivity and in crop cover. Within the context of the USLE (Wischmeier and Smith, 1978), this interaction is quantified by the *C* factor. The *C* factor represents the cumulative product of two distinct factors: the soil loss ratio (*SLR*), which quantifies the instantaneous protective effect of a crop relative to a long-term bare fallow soil in seedbed condition, and the relative erosion index (*REI*), which quantifies the long-term mean seasonal pattern of rain erosivity relative to its annual mean (Wischmeier, 1960):

$$C = \sum SLR \bullet REI \quad (3)$$

This calculation is done incrementally, where increments may be crop stages, months, or days. Typically, the *C* factor is then averaged over a crop rotation to obtain the annual mean value. Accounting for the complete rotation is necessary to encompass the periods between the main crops, which cannot be attributed to a single crop but vary based on the crop sequence.

The *SLR* is close to 1 at seedbed conditions and close to zero for a complete cover. By definition, the *REI* rises from 0 at the start of the year to 1 at the end of the year. Three distinct types of seasonal distribution of the *REI* can be identified (Wischmeier and Smith, 1978; Fig. 4, left panel). The most common is the continental type, characterized by erosive rains primarily occurring during summer months, leading to a sharp increase in the cumulative erosion index during summer. This behavior typically can be found in continental, temperate

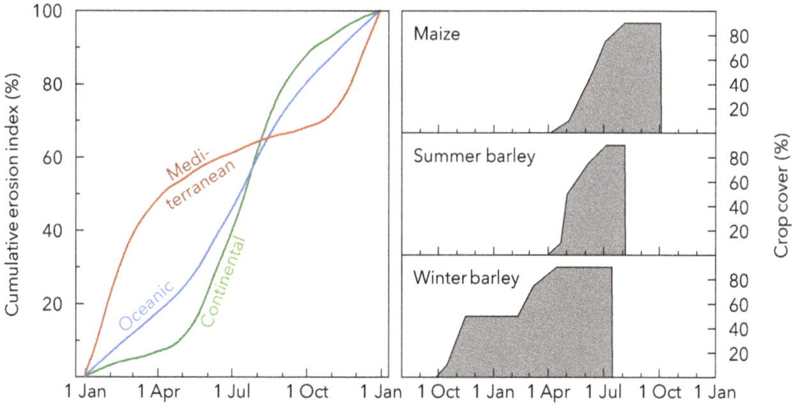

Figure 4 The *C* factor is influenced by the temporal interaction of the seasonal variation of the relative erosion index and soil cover. Left panel: Three types of the relative erosion index result for Mediterranean, oceanic, and continental climates on the northern hemisphere (based on Wischmeier and Smith, 1978). Right panels: Three types of crop development result from fall-sown crops like winter barley (*Hordeum vulgare* L.), spring-sown small-grain crops like summer barley, and spring-sown row crops like maize (*Zea mays* L.). Data on crop cover were taken from Auerswald et al. (2021) and apply to Germany and the second decade of the twenty-first century.

areas like in many states of the mid-west in the USA (Wischmeier and Smith, 1978) or in central Europe (Auerswald et al., 2019a) or northern China (Xie et al., 2016). Conversely, Mediterranean climates experience predominantly erosive rains in winter, while summers are relatively dry. Under an oceanic climate, the seasonality of the *REI* is less pronounced, and erosive rains of similar strength can occur throughout the year.

The seasonal course of *SLR*s predominantly depends on the crop type and can be influenced by erosion control measures such as cover crops, intercropping, reduced tillage, and other practices. We will focus on the effects of crop selection, while the protective measures will not be discussed here as they are based on management decisions only indirectly affected by climate change. In the temporal progression of plant cover, again three types can be discerned (Fig. 4, right panels). Fall-sown crops like winter wheat (*Triticum aestivum* L.) or canola (*Brassica napus* L.) already provide some soil cover during winter but are harvested relatively early. Spring-sown small grains such as summer wheat or summer oats (*Avena sativa* L.) are planted in early spring and harvested shortly after the fall-sown crops. Row crops such as maize (*Zea mays* L.), potatoes (*Solanum tuberosum* L.), or sugar beets (*Beta vulgaris* L.) are sown later in spring and develop soil cover at a slower rate than small grains due to the wider spacing between rows (e.g. maize often has a row spacing of approximately 75 cm compared to small grains with a spacing of around 12 cm). Row crops are usually harvested late in autumn. A late sowing date is especially applied for row crops using C4 photosynthesis like maize or sorghum (*Sorghum bicolor* (L.) Moench), due to the high-temperature sensitivity of the C4 photosynthesis (Ehleringer and Björkman, 1977) that results in a preference of these crops for higher temperatures to germinate and develop.

In general, climate change will alter both the *REI* and the seasonal change in *SLR*s. Given the regional differences in erosion indices and *SLR*s for different crops, the combined effect upon the *C* factor can exert highly variable climate change effects. The subsequent sections will illustrate these effects by focusing on major crops and different climate types. To illustrate the influence of a change in the *REI* or in crop development, all other parameters except for that under focus will not be changed.

3.2 Effects of temporal shifts in soil loss ratios

Elevated temperatures induce accelerated physiological and phenological advancements in all plant species, including crops, which also leads to adapted agricultural field management, especially a change in sowing dates (Menzel et al., 2020). Changes in crop development and phenological stages are particularly pronounced for C4 crops, owing to their large temperature sensitivity. For instance, the date of sowing and germination of maize in Germany has shifted by about 3 weeks between 1960 and 2020 (Fig. 5). Similar effects have also been found in the USA (Sacks and Kucharik, 2011). In comparison, C3 crops are less affected because their physiological development primarily is governed by the length of the light period (Olesen et al., 2012), which remains unaffected by climate change. Nevertheless, in a warmer climate, sowing becomes

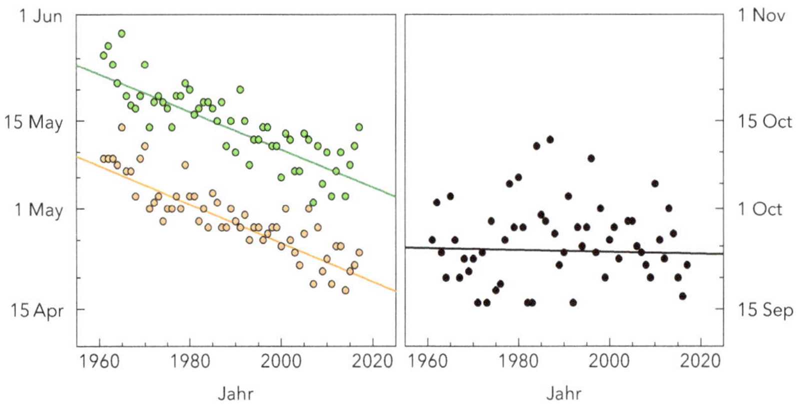

Figure 5 Change in the date of sowing (orange), germination (green), and harvest (black) of maize (*Zea mays* L.) in Germany over six decades. Data were taken from Auerswald and Menzel (2021); each data point is the median of more than 1500 locations more or less equally distributed throughout Germany. Lines are linear regressions (very highly significant in the left panel and insignificant in the right panel).

delayed for fall-sown crops, mainly for phytosanitary reasons, whereas spring-sown small grains may be planted earlier, mainly to make better use of winter moisture (Menzel et al., 2020; Auerswald et al., 2021). Olesen et al. (2012) could show for two climate model projections under the A1B emission scenario (representing a mid-range scenario for greenhouse gas emissions, as defined by the Intergovernmental Panel on Climate Change in 2000; Nakićenović and Swart, 2000) that the advancement in phenological dates of small grain and maize will continue by 1–3 weeks until 2040, being larger for maize and depending on the used climate model and the modeled region within Europe.

The consequences of such shifts exhibit considerable variation, contingent upon the specific *REI* and crop type, even if all other conditions remain unchanged. To reduce complexity, we illustrate this for the hypothetical cases of a fall-sown small grain (winter rye, *Secale cereale* L.) and a spring-sown row crop (maize), which both can be grown continuously, thereby fulfilling the requirements of the *C* factor of covering a complete rotation including the periods between subsequent main crops. Combining the unchanged relative erosion indices for continental, oceanic, and Mediterranean climates as given in Fig. 4 with the unchanged *SLR*s by Auerswald et al. (2021) but assuming a 10-day delay in the fall sowing of winter rye, coupled with a 10-day advancement in crop development during spring and summer due to rising temperatures, the *C* factor of continuous rye decreases in continental climates (Fig. 6, right panel). This reduction is primarily attributed to improved soil cover during the spring and early summer months. In contrast, under Mediterranean climates, the same shift in crop development increases the *C* factor of rye as the critical period between seedbed and 50% soil cover aligns with the time when most erosive rains occur in this climate. Conversely, the changes in the *C*

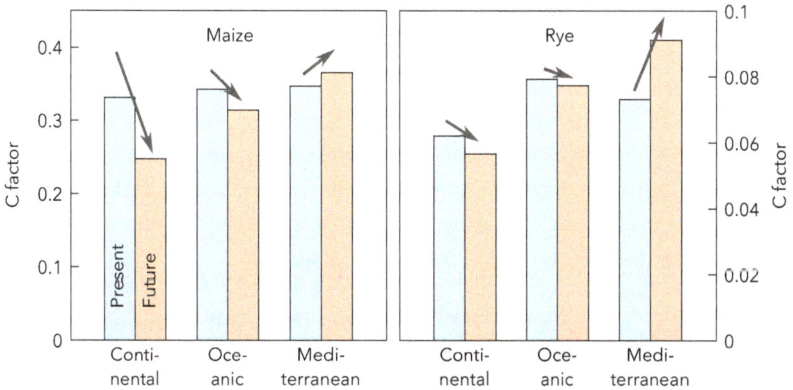

Figure 6 Influence of climate-change-induced shifts in the phenological development of maize and winter rye on the *C* factor assuming an unchanged continental, oceanic, and Mediterranean distribution of the relative erosion index. Soil loss ratios were taken from Auerswald et al. (2021) and assumed to be independent of climate and climate change.

factor are minimal under oceanic climates. Given the nearly even distribution of erosivity throughout the year, the date of the critical seedbed period becomes inconsequential. Although a shorter duration of this period lowers the *C* factor, the change remains marginal. In cases where the *REI* is evenly distributed, only 0.3% of the annual erosivity will occur per day. Consequently, a shift of a few days has little impact. The effect on soil loss is even smaller, as other periods must then become longer to complete a full year.

Assuming an accelerated growth of maize by one month significantly decreases the *C* factor of continuous maize in continental climates (by 25%), as crop development will already be more advanced during the onset of the primary erosive-event period (Fig. 6, left panel). Under oceanic climates, where the peak of rain erosivity in summer is minor, this effect will be small like it was for rye. Conversely, in Mediterranean climates, the opposite holds true. The *C* factor will increase due to earlier sowing, which shifts the critical state to the winter half of the year when the majority of the *REI* is expected.

In summary, accelerated crop growth will lower the *C* factor for both rye and maize under continental climates, while it will increase for both crops under Mediterranean climates. Changes will be inconsequential for both crops under oceanic climates. It is important to note that these crops serve as mere examples and cannot be generalized. The changes may be significantly different for other crops. Also, the assumption of identical *SLR*s and crop stage dates is unlikely and only serves the purpose of illustrating the influence of changes in phenological development.

More pronounced alterations will occur in regions where rising temperatures enable the cultivation of temperature-sensitive crops like maize that previously could not be grown (Kucharik and Serbin, 2008). This phenomenon is also evident in Germany, where the spread of maize starting in the 1960s was slower in northern regions compared to southern regions, despite identical socio-economic conditions (Fig. 7, left panel). This difference agreed with the consistently colder temperatures (about 0.4 °C; DWD, 2023) during maize germination in May in northern Germany and disappeared gradually with increasing temperatures due to climate change. Thus, the difference between north and south disappears when the decadal mean maize fraction between 1960 and 2019 is plotted versus the decadal mean May temperature, indicating that the higher temperatures can provide new options for farmers (Fig. 7, right panel). Given that conventionally grown maize exhibits one of the highest *C* factors (Ghidey and Alberts, 1998; Auerswald et al., 2021), the expansion of maize cultivation has intensified erosion.

Soybean (*Glycine max* L. Merrill) is another erosion-prone crop (Ghidey and Alberts, 1998; Basic et al., 2004), which was absent in many areas due to an unfavorable climate that has started to spread enormously with increasing temperatures (Auerswald et al., 2021) and is predicted to spread further with

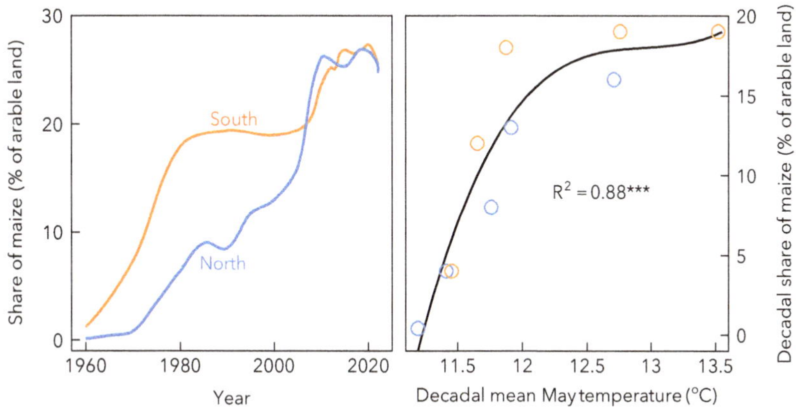

Figure 7 Temporal development in the fraction of maize on arable land in south Germany (federal state of Bavaria) and north Germany (federal state of Schleswig-Holstein) between 1960 and 2022 (left panel) and correlation of the decadal mean fractions of maize with the decadal mean air temperatures in May for both regions (right panel; data taken from DWD, 2023).

climate change (O'Neal et al., 2005). Other erosion-prone warm-season crops that were absent in former decades also gradually increased their acreage very highly significantly during the last decade (2012 and 2022) although they still remain on a low share of arable land. For example in the German federal state of Bavaria grain sorghum (*Sorghum bicolor* L. Moench), proso millet (*Panicum miliaceum* L.), and sweet potato (*Impomoea batatas* L. Lam.) from being absent before 2000, they ranked no. 31, 47, and 118 out of 164 grown crops in 2019 (Fig. 8).

The gradual transition to warm-season crops in response to climate warming entails two significant implications, both of which can be elucidated using these data. Firstly, warm-season crops contribute more to soil erosion than other crops. The summable *C* factors called gammas as developed by Auerswald et al. (2021) for calculating an average *C* factor based on cropping information, reveal that all aforementioned warm-season crops exhibit high gamma values: maize 0.252, amaranth (*Aramanth* species) 0.245, soybean 0.241, sorghum and millet 0.148. In contrast, the average *C* factor for crop rotations, determined using the gamma approach, was 0.132 for Germany in 2016. Consequently, substituting crops according to their average share with any of the aforementioned warm-season crops would invariably increase the average *C* factor due to their gamma values exceeding the average. Moreover, if the substitution involved replacing small grains (with an average gamma of all fall-sown small-grain crops at 0.077), the increase in the average *C* factor would be even more substantial.

For sweet potato, quinoa, and upland rice (*Oryza sativa* L.), Auerswald et al. (2021) did not provide specific gamma values, despite listing 57 crops

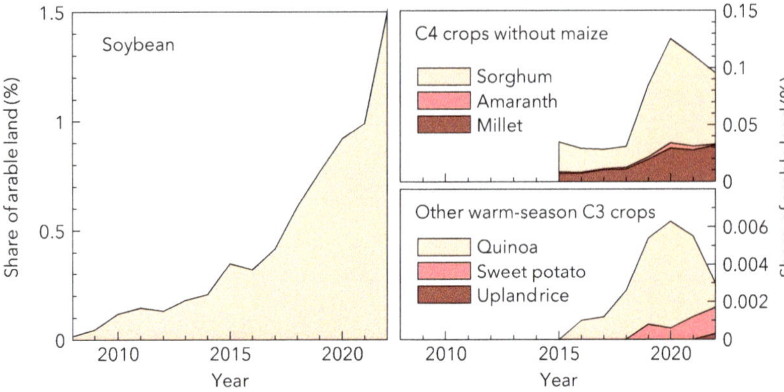

Figure 8 Warming-induced increase in warm-season crops in the federal state of Bavaria in Germany (data were taken from the integrated administration and control system for European Community aid schemes; they are unpublished and confidential and include all arable areas larger than 200 m^2).

and crop management options. Consequently, there is a lack of knowledge regarding their erosion behavior under regional agricultural conditions. Hence, the second crucial implication arising from the shift in crops is that the quantification of their impact on erosion is unfeasible until they are cultivated on a significant portion of land, allowing for the estimation of their erosion behavior, or until suitable experiments are conducted to assess the *C* factor of rotations incorporating these new crops. Both approaches necessitate a considerable span of time during which these crops may substantially expand their share of arable land, while their erosion effects cannot be adequately evaluated in regulatory or administrative frameworks.

Ironically, society promotes the cultivation of maize and soybean as a means to mitigate climate change through the production of biofuel and biogas (Cavalett and Ortega, 2010; Xu et al., 2013; Levin et al., 2021), despite the fact that this practice places additional pressure on the soil resource. Thus, although the rising erosivity calls for the implementation of soil and water conservation strategies, it appears that such initiatives still fall short, and instead, a trend toward land use practices that result in increased erosion can be observed.

Furthermore, different crop management practices may become viable. Double cropping, involving the sequential cultivation of two crops within a single year, may become feasible in certain regions due to an extended vegetation period (Meza et al., 2008; Seifert and Lobell, 2015). This entails preparing a seedbed for the second crop during the summer, coinciding with the peak of the *REI* in continental climates. In Mediterranean climates, the likelihood of the double cropping effect will be less pronounced in rainfed agriculture due to lower rainfall during the summer months.

Irrigation may be applied to mitigate climate change-induced prolonged dry periods. This might substantially increase erosion rates under sprinkler irrigation (King and Bjorneberg, 2011; Batista et al., 2023), furrow irrigation (Trout, 1996), and even drip irrigation (Cerdà et al., 2021). This additional erosion can occur due to the additional drop and flow forces exerted by the irrigation water itself but also erosion by subsequent natural rains can be enhanced due to the change in soil properties (Wallach et al., 2005).

3.3 Influences of the shift in the relative erosion index

The protective effect of crop cover is not only affected by changes in plant development but also via potential climate change-induced shifts in the seasonality of rainfall erosivity. It is important to note that a change in seasonality is independent from the general increase in the annual *R* factor. While a general increase of erosivity can be expected for all three types of the *REI* (Fig. 4, left), it can be anticipated particularly under a continental climate, that rain erosivity will experience a notable relative increase particularly during the winter season, as precipitation transitions from snowfall to rainfall. Furthermore, the likelihood of convective rains, which were previously rare during the winter half-year, is heightened. Consequently, the distribution of the *REI* becomes more evenly spread throughout the year. This phenomenon has already been demonstrated in studies conducted in China (Xu et al., 2023) and Germany (Auerswald et al., 2019a). In contrast, less pronounced effects are expected in oceanic and Mediterranean climates, as the most significant impact stems from the substitution of snowfall with rainfall, which generally occurs only at higher altitudes in oceanic and Mediterranean climates. However, seasonal changes are also reported for Mediterranean areas (Diodato and Bellocchi, 2009; Vallebona et al., 2015) but a comprehensive analysis of the changes in the *REI* is currently lacking, in particular for oceanic climates.

The increase in the *REI* during the winter months, leading to a corresponding reduction in the *REI* during the summer months (Fig. 9, left panel), primarily impacts crops whose vulnerable stages of seedbed and early growth occur during the winter half of the year. This includes fall-sown crops and early in spring-sown small grains (Fig. 9, right panel). All species of spring-sown small grain are sown around the same time and exhibit similar developmental patterns thereafter. Consequently, the alteration in the *REI* affects them uniformly, as evidenced by the small standard deviation in Fig. 9. The seemingly minor change in the *REI* causes the *C* factor for these crops to increase by almost 100%. Consequently, erosion of these crops increases twofold due to climate change: due to the increase in *R* and due to the increase in *C*. Furthermore, the difference in *C* between the spring-sown small grains and row crops, which are sown later in spring, diminishes as the *C* factor for row crops undergoes only

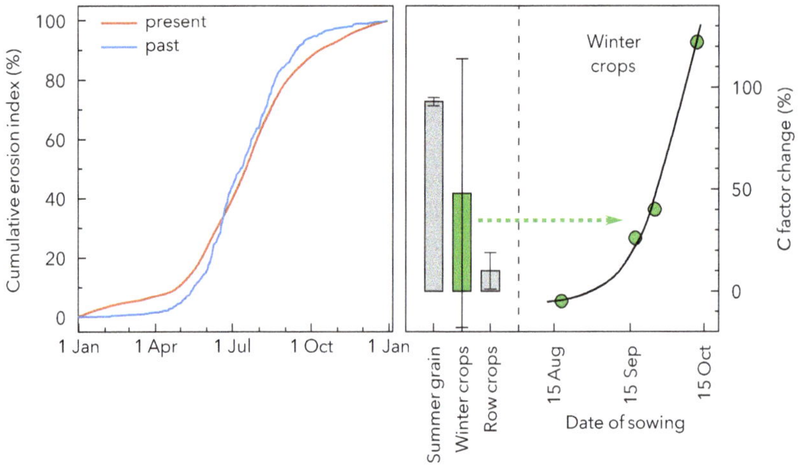

Figure 9 Influence of a higher proportion of erosivity during winter on the *C* factors for a continental distribution of erosivity (data were taken from Auerswald and Menzel, 2021). Left panel: Cumulative relative erosion index in the 1960s (past) and in the first two decades of the twenty-first century (present) in Germany. Right panel: Resulting change of the *C* factor of spring-sown row crops, fall-sown winter crops, and early spring-sown summer grains (bars; the error bars indicate the standard deviation between different crops); *C* factor change of winter crops depending on the typical date of sowing (circles; the data points show from left to right; canola, winter barley, winter rye, and winter wheat).

minimal changes (Fig. 9, left panel). The changes during the winter and summer halves of the year nearly offset one another for row crops.

The situation for fall-sown winter crops is more variable. Early-sown winter crops like canola establish substantial plant cover before winter, rendering them unaffected by the rise in the relative erosion index during winter (Fig. 9, right panel). Conversely, crops such as winter wheat, sown late in the season, exhibit minimal cover during winter and are consequently more heavily impacted. In the example presented in Fig. 9, the *C* factor for winter wheat increases by 125%. As a result of the multiplicative nature of the USLE, this implies that soil erosion associated with winter wheat has increased by 350% (2.25 × 2 - 1) under consideration that the *R* factor also has doubled due to climate change.

4 Snowmelt erosion

The increase in winter rain at the expense of snowfall that increased the erosion index during winter months under continental climate may lead to the expectation that the increased soil loss by erosive rains during winter may be compensated by lower soil loss due to snowmelt. This may be the case. Soil loss by snowmelt may be even more lowered in areas where vegetation growth is strongly restricted by low temperatures like at high altitudes or in the tundra and taiga zones. Better growth due to a warming climate may additionally lower soil loss (Wang et al., 2023). This line of thought is used to explain why soil erosion has declined after the 1980s in the forest and forest-steppe zones of the Russian Plain while it has increased in the steppe zone where the increase in rain erosivity dominated (Golosov et al., 2018).

The net effect of an increase in erosion due to more rain erosivity during winter and the decrease in snowmelt erosion will vary and is difficult to predict. This uncertainty is largely due to the fact that our quantitative knowledge of how much snowmelt erosion can be expected is considerably less developed than our understanding of rainfall-induced erosion (Milazzo et al., 2023).

However, climate change must not necessarily decrease snowmelt erosion. The opposite, an increase in snowmelt erosion as an effect of climate change, is also reported. Two mechanisms contribute. First, an increase in winter precipitation can cause more snow to accumulate. Second, warm and intense rain may become more likely to fall on the snowpack. This combination of a warm rain falling on snow and causing a rapid melting eventually coupled with frozen soil underneath the snow has been reported to cause the most severe snowmelt erosion events (Harr, 1981; Zuzel et al., 1982). For Norway, it is expected that these conditions may happen more often as a consequence of climate change (Deelstra et al., 2011).

5 Conclusion

Climate change has multiple influences on soil erosion, most of them leading to an amplification of erosion by water. Furthermore, it invalidates scientific knowledge because values, which formerly could be treated as site-specific constants now have become variables and need to be re-assessed regularly. In particular, this is true for rain erosivity. Although an increase in precipitation of only 2–3% per one-degree increase in temperature is expected, the increase in rainfall erosivity with increasing temperature is much larger. This results from several amplifying processes: (i) an increase in rainfall amount in case of erosive precipitation events, (ii) an increase in precipitation intensity and droplet size for these events, (iii) an accumulation of erosive events, and (vi) an accumulation of situations with lower wind speeds and thus potentially stationary heavy precipitation events. The combined impact of these processes was demonstrated through an analysis of rainfall erosivity data from Germany, revealing a significant doubling of erosivity over the past six decades. Climate modeling indicated that the trend will continue in the coming decades under realistic emission scenarios. Nevertheless, it is crucial to acknowledge that a considerable spatial variability in trend may exist, although there is compelling evidence that elevated temperatures will generally contribute to heightened erosivity.

In addition to changes in annual erosivity, a change in the seasonal distribution of erosivity and in soil cover by arable crops can also be observed.

Both impact the protective power of crops against erosion as quantified in the *C* factor. The *C* factor change depends on the crops and on the climatic region, which results in a large diversity in the individual changes. If only considering the change in crop growth (physiology and phenology) and the associated adoption in agricultural management practice (e.g. sowing earlier in spring), trends differ between crop types like row crops or small grain and between different climatic regions exhibiting continental, oceanic, or Mediterranean distribution of erosivity. For example, while an increase in the *C* factor for maize is expected for the Mediterranean climate, the opposite is the case for continental and oceanic climates. Even more pronounced climate-induced effects might result from shifts in the area cultivated with different crops. In particular, warm-season row crops, which usually facilitate erosion, may expand. For C4 plants (such as the erosion-prone maize), which benefit particularly from the temperature increase in temperate climates, a further expansion of the cultivated area to somewhat colder regions can be observed. Hence, the relative regional *C* factor might increase, even if the *C* factor of maize itself decreases. Apart from these general climate change-induced shifts in seasonality of crop cover in different climatic regions, its overall effect upon the *C* factor can only be analyzed in combination with potential seasonal shifts in rainfall erosivity. Under continental climate, where a shift of the *REI* into the winter half-year is expected due to an increase of rainfall at the expense of snowfall, this will potentially increase the *C* factor of winter crops and early sown summer grains.

In cases where climate change not only induces temperature rise but also exacerbates drought conditions, the potential for erosive processes to intensify becomes even more pronounced. The heightened occurrence of wildfires that remove the protective vegetation and litter cover can further aggravate erosion and the implementation of irrigation to alleviate drought may also inadvertently contribute to additional erosion.

6 Appendix: Euro-Cordex climate models used for the characterization of future erosivity

Number	Global climate model (GCM)	Regional climate model (RCM)	Driving ensemble member
4	EC-EARTH	KNMI-RACMO22E	r1i1p1
6	EC-EARTH	CCLM 4-8-17	r12i1p1
7	EC-EARTH	SMHI-RCA4	r12i1p1
9	IPSL-CM5A-MR	SMHI-RCA4	r1i1p1
12	HadGEM2-ES	SMHI-RCA4	r1i1p1
13	MPI-ESM-LR	CCLM 4-8-17	r1i1p1
15	MPI-ESM-LR	SMHI-RCA4	r1i1p1

7 References

Arnoldus, H. M. J. 1977. Assessing soil degradation: Methodology used to determine the maximum average soil loss due to sheet and rill erosion in Morocco. FAO Soil Bulletin 34, Rome. Available at: https://www.fao.org/3/ar114e/ar114e.pdf [last access: 29 May 2023].

Auerswald, K. and Menzel, A. 2021. Change in erosion potential of crops due to climate change. *Agricultural and Forest Meteorology* 300, 108338. doi:10.1016/j.agrformet.2021.108338.

Auerswald, K., Fischer, F. K., Winterrath, T. and Brandhuber, R. 2019a. Rain erosivity map for Germany derived from contiguous radar rain data. *Hydrology and Earth System Sciences* 23(4), 1819–1832. doi:10.5194/hess-23-1819-2019.

Auerswald, K., Fischer, F., Winterrath, T., Elhaus, D., Maier, H. and Brandhuber, R. 2019b. Klimabedingte Veränderung der Regenerosivität seit 1960 und Konsequenzen für Bodenabtragsschätzungen. In: Bachmann, G., König, W. and Utermann, J. (Eds). *Bodenschutz, Ergänzbares Handbuch der Maßnahmen und Empfehlungen für Schutz, Pflege und Sanierung von Böden, Landschaft und Grundwasser (Loseblatt-sammlung)*, Erich Schmidt Verlag, Berlin, 4090, 21 p.

Auerswald, K., Fischer, F. and Winterrath, T. 2019c. R-Faktor – Regenerosivität. In: *Bayerisches Landesamt für Umwelt (Hrsg.): Pilotstudie, "Klimawirkungskarten Bayern"*. UmweltSpezial, Bayerisches Landesamt für Umwelt, Augsburg, 61–69.

Auerswald, K., Ebertseder, F., Levin, K., Yuan, Y., Prasuhn, V., Plambeck, N. O., Menzel, A. and Kainz, M. 2021. Summable C factors for contemporary soil use. *Soil and Tillage Research* 213, 105155. doi:10.1016/j.still.2021.105155.

Basic, F., Kisic, I., Mesic, M., Nestroy, O. and Butorac, A. 2004. Tillage and crop management effects on soil erosion in central Croatia. *Soil and Tillage Research* 78(2), 197–206.

Batista, P. V. G., Baptista, V. B. D. S., Wilken, F., Seufferheld, K., Quinton, J. N. and Fiener, P. 2023. First evidence of widespread, severe soil erosion underneath centre-pivot irrigation systems. *Science of the Total Environment* 888, 164119. doi:10.1016/j.scitotenv.2023.164119.

Biasutti, M. and Seager, R. 2015. Projected changes in US rainfall erosivity. *Hydrology and Earth System Sciences* 19(6), 2945–2961.

Bürger, G., Heistermann, M. and Bronstert, A. 2014. Towards subdaily rainfall disaggregation via Clausius-Clapeyron. *Journal of Hydrometeorology* 15(3), 1303–1311. doi:10.1175/JHM-D-13-0161.1.

Cavalett, O. and Ortega, E. 2010. Integrated environmental assessment of biodiesel production from soybean in Brazil. *Journal of Cleaner Production* 18(1), 55–70.

Cerdà, A., Novara, A. and Moradi, E. 2021. Long-term non-sustainable soil erosion rates and soil compaction in drip-irrigated citrus plantation in eastern Iberian Peninsula. *Science of the Total Environment* 787, 147549.

Chapman, S., Birch, C. E., Galdos, M. V., Pope, E., Davie, J., Bradshaw, C., Eze, S. and Marsham, J. H. 2021. Assessing the impact of climate change on soil erosion in East Africa using a convection-permitting climate model. *Environmental Research Letters* 16(8), 084006. doi:10.1088/1748-9326/ac10e1.

Chen, W., Huang, Y.-C., Lebar, K. and Bezak, N. 2023. A systematic review of the incorrect use of an empirical equation for the estimation of the rainfall erosivity around the globe. *Earth-Science Reviews* 238, 104339.

Coumou, D., Petoukhov, V., Rahmstorf, S., Petri, S. and Schellnhuber, H. J. 2014. Quasi-resonant circulation regimes and hemispheric synchronization of extreme weather in boreal summer. *Proceedings of the National Academy of Sciences* 111(34), 12331–12336.

Coumou, D., Lehmann, J. and Beckmann, J. 2015. The weakening summer circulation in the northern hemisphere mid-latitudes. *Science* 348(6232), 324–327. doi:10.1126/science.1261768.

Crossman, J., Futter, M. N., Oni, S. K., Whitehead, P. G., Jin, L., Butterfield, D., Baulch, H. M. and Dillon, P. J. 2013. Impacts of climate change on hydrology and water quality: future proofing management strategies in the Lake Simcoe watershed, Canada. *Journal of Great Lakes Research* 39(1), 19–32.

Deelstra, J., Øygarden, L., Blankenberg, A.-G. B. and Eggestad, H. O. 2011. Climate change and runoff from agricultural catchments in Norway. *International Journal of Climate Change Strategies and Management* 3(4), 345–360.

Deumlich, D. and Gericke, A. 2020. Frequency trend analysis of heavy rainfall days for Germany. *Water* 12(7), 1950.

Diodato, N. and Bellocchi, G. 2009. Assessing and modelling changes in rainfall erosivity at different climate scales. *Earth Surface Processes and Landforms* 34(7), 969–980.

Diodato, N., Borrelli, P., Fiener, P., Bellocchi, G. and Romano, N. 2017. Discovering historical rainfall erosivity with a parsimonious approach: A case study in Western Germany. *Journal of Hydrology* 544, 1–9.

Dore, M. H. I. 2005. Climate change and changes in global precipitation patterns: what do we know? *Environment International* 31(8), 1167–1181.

DWD 2023. Zeitreihen fuer Gebietsmittel fuer Bundeslaender und Kombinationen von Bundeslaender. Available at: https://opendata.dwd.de/climate_environment/CDC/regional_averages_DE/monthly/air_temperature_mean/regional_averages_tm_05.txt [last access: 10 June 2023].

Ehleringer, J. R. and Björkman, O. 1977. Quantum yields for CO2 uptake in C3 and C4 plants - dependence on temperature, CO2, and O2 concentration. *Plant Physiology* 59(1), 86–90.

Fiener, P. and Auerswald, K. 2009. Spatial variability of rainfall on a sub-kilometre scale. *Earth Surface Processes and Landforms* 34(6), 848–859. doi:10.1002/esp.1779.

Fiener, P., Neuhaus, P. and Botschek, J. 2013. Long-term trends in rainfall erosivity - analysis of high resolution precipitation time series (1937–2007) from Western Germany. *Agricultural and Forest Meteorology* 171–172, 115–123.

Fischer, F., Hauck, J., Brandhuber, R., Weigl, E., Maier, H. and Auerswald, K. 2016. Spatio-temporal variability of erosivity estimated from highly resolved and adjusted radar rain data (RADOLAN). *Agricultural and Forest Meteorology* 223, 72–80. doi:10.1016/j.agrformet.2016.03.024.

Fischer, F. K., Winterrath, T. and Auerswald, K. 2018. Temporal- and spatial-scale and positional effects on rain erosivity derived from point-scale and contiguous rain data. *Hydrology and Earth System Sciences* 22(12), 6505–6518. doi:10.5194/hess-22-6505-2018.

Golosov, V., Yermolaev, O., Litvin, L., Chizhikova, N., Kiryukhina, Z. and Safina, G. 2018. Influence of climate and land use changes on recent trends of soil erosion rates within the Russian Plain. *Land Degradation and Development* 29(8), 2658–2667.

Harr, R. D. 1981. Some characteristics and consequences of snowmelt during rainfall in western Oregon. *Journal of Hydrology* 53(3–4), 277–304.

Hübener, H., Bülow, K., Fooken, C., Früh, B., Hoffmann, P., Höpp, S., Keuler, K., Menz, C., Mohr, V., Radtke, K., Ramthun, H., Spekat, A., Steger, C., Toussaint, F., Warrach-Sagi, K. and Woldt, M. 2017. *ReKliEs-De Regionale Klimaprojektionen Ensemble für Deutschland*. doi:10.2312/WDCC/ReKliEsDe_Ergebnisbericht.

Ghidey, F. and Alberts, E. E. 1998. Runoff and soil losses as affected by corn and soybean tillage systems. *Journal of Soil and Water Conservation* 53, 64-70.

Greuell, W. 2000. Melt-water accumulation on the surface of the Greenland Ice Sheet: effect on albedo and mass balance. *Geografiska Annaler: Series A, Physical Geography* 82(4), 489-498. doi:10.1111/j.0435-3676.2000.00136.x.

Haerter, J. O. and Berg, P. 2009. Unexpected rise in extreme precipitation caused by a shift in rain type? *Nature Geoscience* 2(6), 372-373.

Jones, B. K., Saylor, J. R. and Testik, F. Y. 2010. Raindrop morphodynamics. In: Testik, F. Y., Gebremichael, M. (Eds). *Rainfall: State of the Science*. doi:10.1029/2009GM000928.

Jones, C., Giorgi, F. and Asrar, G. 2011. The coordinated regional downscaling experiment: CORDEX, an international downscaling link to CMIP5. *Clivar Exchanges* 16, 34-40.

King, B. A. and Bjorneberg, D. L. 2011. Evaluation of potential runoff and erosion of four center pivot irrigation sprinklers. *Applied Engineering in Agriculture* 27(1), 75-85. doi:10.13031/2013.36226.

Kinnell, P. I. A. 1991. The effect of flow depth on sediment transport induced by raindrops impacting shallow flows. *Transactions of the American Society of Agricultural Engineers* 34, 161-166.

Knist, S., Goergen, K., Buonomo, E., Christensen, O. B., Colette, A., Cardoso, R. M., Fealy, R., Fernández, J., García-Díez, M., Jacob, D., Kartsios, S., Katragkou, E., Keuler, K., Mayer, S., van Meijgaard, E., Nikulin, G., Soares, P. M. M., Sobolowski, S., Szepszo, G., Teichmann, C., Vautard, R., Warrach-Sagi, K., Wulfmeyer, V. and Simmer, C. 2017. Land-atmosphere coupling in EURO-CORDEX evaluation experiments. *Journal of Geophysical Research: Atmospheres* 122(1), 79-103. doi:10.1002/2016JD025476.

Kucharik, C. J. and Serbin, S. P. 2008. Impacts of recent climate change on Wisconsin corn and soybean yield trends. *Environmental Research Letters* 3(3), 034003.

Laws, J. O. and Parsons, D. A. 1943. The relation of raindrop size to intensity. *Transactions of the American Geophysical Union* 24(2), 452-460. doi:10.1029/TR024i002p00452.

Lenderink, G. and van Meijgaard, E. 2008. Increase in hourly precipitation extremes beyond expectations from temperature changes. *Nature Geoscience* 1(8), 511-514, doi:10.1038/ngeo262.

Levin, K. S., Auerswald, K., Reents, H. J. and Hülsbergen, K.-J. 2021. Effects of energy crop rotations and digestate fertilisation on soil organic carbon in the topsoil in an organic field experiment. *Agronomy* 11(7), 1393. doi:10.3390/agronomy11071393.

Lobo, D., Lozano, Z. and Delgado, F. 2005. Water erosion risk assessment and impact on productivity of a Venezuelan soil. *CATENA* 64(2-3), 297-306.

Lochbihler, K., Lenderink, G. and Siebesma, A. P. 2017. The spatial extent of rainfall events and its relation to precipitation scaling. *Geophysical Research Letters* 44(16), 8629-8636. doi:10.1002/2017GL074857.

Loriaux, J. M., Lenderink, G., De Roode, S. R. and Siebesma, A. P. 2013. Understanding convective extreme precipitation scaling using observations and an entraining plume model. *Journal of the Atmospheric Sciences* 70(11), 3641-3655, doi:10.1175/JAS-D-12-0317.1.

Mann, M. E., Rahmstorf, S., Kornhuber, K., Steinman, B. A., Miller, S. K., Petri, S. and Coumou, D. 2018. Projected changes in persistent extreme summer weather events: the role of quasi-resonant amplification. *Science Advances* 4(10), eaat3272.

Menzel, A., Yuan, Y., Matiu, M., Sparks, T., Scheifinger, H., Gehrig, R. and Estrella, N. 2020. Climate change fingerprints in recent European plant phenology. *Global Change Biology* 26, 2599–2612. doi:10.1111/gcb.15000.

Met Office Hadley Centre 2023. Deviation from global mean surface temperature 1850 to 1900. HadCRUT.5.0.1.0 model, median of 200 calculated time series. Climate Research Unit. Available at: https://www.umweltbundesamt.de/daten/klima/trends-der-lufttemperatur#steigende-durchschnittstemperaturen-weltweit [last access: 15 May 2023].

Meza, F. J., Silva, D. and Vigil, H. 2008. Climate change impacts on irrigated maize in Mediterranean climates: evaluation of double cropping as an emerging adaptation alternative. *Agricultural Systems* 98(1), 21–30.

Milazzo, F., Francksen, R. M., Zavattaro, L., Abdalla, M., Hejduk, S., Enri, S. R., Pittarello, M., Price, P. N., Schils, R. L. M., Smith, P. and Vanwalleghem, T. 2023. The role of grassland for erosion and flood mitigation in Europe: A meta-analysis. *Agriculture, Ecosystems and Environment* 348, 108443.

Mohr, S. and Kunz, M. 2013. Recent trends and variabilities of convective parameters relevant for hail events in Germany and Europe. *Atmospheric Research* 123, 211–228.

Mohr, S., Ehret, U., Kunz, M., Ludwig, P., Caldas-Alvarez, A., Daniell, J. E., Ehmele, F., Feldmann, H., Franca, M. J., Gattke, C., Hundhausen, M., Knippertz, P., Küpfer, K., Mühr, B., Pinto, J. G., Quinting, J., Schäfer, A. M., Scheibel, M., Seidel, F. and Wisotzky, C. 2023. A multi-disciplinary analysis of the exceptional flood event of July 2021 in central Europe -- Part 1: Event description and analysis. *Natural Hazards and Earth System Sciences* 23(2), 525-551. doi:10.5194/nhess-23-525-2023.

Mueller, E. N. and Pfister, A. 2011. Increasing occurrence of high-intensity rainstorm events relevant for the generation of soil erosion in a temperate lowland region in central Europe. *Journal of Hydrology* 411(3–4), 266–278, doi:10.1016/j.jhydrol.2011.10.005.

Nakićenović, N. and Swart, R. (Eds) 2000. *Special Report on Emissions Scenarios: A Special Report of Working Group III of the Intergovernmental Panel on Climate Change*, Cambridge University Press, ISBN 0-521-80081-1.

Odemark, K., Müller, M., Palerme, C. and Tveito, O. E. 2023. Recent changes in circulation patterns and their opposing impact on extreme precipitation at the west coast of Norway. *Weather and Climate Extremes* 39, 100530. doi:10.1016/j.wace.2022.100530.

Olesen, J. E., Børgesen, C. D., Elsgaard, L., Palosuo, T., Rötte, R. P., Skjelvåg, A. O., Peltonen-Sainio, P., Börjesson, T., Trnka, M., Ewert, F., Siebert, S., Brisson, N., Eitzinger, J., van Asselt, E. D., Oberforster, M. and van der Fels-Klerx, H. J. 2012. Changes in time of sowing, flowering and maturity of cereals in Europe under climate change. *Food Additives and Contaminants. Part A, Chemistry, Analysis, Control, Exposure and Risk Assessment* 29(10), 1527–1542.

Oliver, J. E. 1980. Monthly precipitation distribution: A comparative index. *The Professional Geographer* 32(3), 300–309.

O'Neal, M. R., Nearing, M. A., Vining, R. C., Southworth, J. and Pfeifer, R. A. 2005. Climate change impacts on soil erosion in Midwest United States with changes in crop management. *CATENA* 61(2–3), 165–184.

Pausas, J. G. and Fernández-Muñoz, S. 2012. Fire regime changes in the western Mediterranean Basin: from fuel-limited to drought-driven fire regime. *Climatic Change* 110(1-2), 215-226.

Pham, T. V., Steger, C., Rockel, B., Keuler, K., Kirchner, I., Mertens, M., Rieger, D., Zängl, G. and Früh, B. 2021. ICON in climate limited-area mode (ICON release version 2.6. 1): A new regional climate model. *Geoscientific Model Development* 14(2), 985-1005. doi:10.5194/gmd-14-985-2021.

Pierce, F. J., Larson, W. E., Dowdy, R. H. and Graham, W. A. P. 1983. Productivity of soils: Assessing long-term changes due to erosion. *Journal of Soil and Water Conservation* 38, 39-44.

Prein, A. F., Langhans, W., Fosser, G., Ferrone, A., Ban, N., Goergen, K., Keller, M., Tölle, M., Gutjahr, O., Feser, F., Brisson, E., Kollet, S., Schmidli, J., van Lipzig, N. P. M. and Leung, R. 2015. A review on regional convection-permitting climate modeling: demonstrations, prospects, and challenges. *Reviews of Geophysics* 53(2), 323-361. doi:10.1002/2014RG000475.

Quinton, J. N., Öttl, L. K. and Fiener, P. 2022. Tillage exacerbates the vulnerability of cereal crops to drought. *Nature Food* 3(6), 472-479.

Rantanen, M., Karpechko, A. Y., Lipponen, A., Nordling, K., Hyvärinen, O., Ruosteenoja, K., Vihma, T. and Laaksonen, A. 2022. The Arctic has warmed nearly four times faster than the globe since 1979. *Communications Earth & Environment* 3(1), 168. doi:10.1038/s43247-022-00498-3.

Rockel, B., Will, A. and Hense, A. 2008. The regional climate model COSMO-CLM (CCLM). *Meteorologische Zeitschrift* 17(4), 347-348, doi:10.1127/0941-2948/2008/0309.

Roderick, M. L., Sun, F., Lim, W. H. and Farquhar, G. D. 2014. A general framework for understanding the response of the water cycle to global warming over land and ocean. *Hydrology and Earth System Sciences* 18(5), 1575-1589.

Rogler, H. and Schwertmann, U. 1981. Erosivität der Niederschläge und Isoerodentkarte Bayerns. *Zeitschrift für Kulturtechnik und Flurbereinigung* 22, 99-112.

Rybka, H., Haller, M., Brienen, S., Brauch, J., Früh, B., Junghänel, T., Lengfeld, K., Walter, A. and Winterrath, T. 2022. Convection-permitting climate simulations with COSMO-CLM for Germany: analysis of present and future daily and subdaily extreme precipitation, development. *Meteorologische Zeitschrift* 64, 65. doi:10.1127/metz/2022/1147.

Sacks, W. J. and Kucharik, C. J. 2011. Crop management and phenology trends in the U. S. Corn Belt: impacts on yields, evapotranspiration and energy balance. *Agricultural and Forest Meteorology* 151(7), 882-894.

Scheffer, M., Carpenter, S., Foley, J. A., Folke, C. and Walker, B. 2001. Catastrophic shifts in ecosystems. *Nature* 413(6856), 591-596.

Seifert, C. A. and Lobell, D. B. 2015. Response of double cropping suitability to climate change in the United States. *Environmental Research Letters* 10(2), 024002.

Slingo, J., Bates, P., Bauer, P., Belcher, S., Palmer, T., Stephens, G., Stevens, B., Stocker, T. and Teutsch, G. 2022. Ambitious partnership needed for reliable climate prediction. *Nature Climate Change* 12(6), 499-503. doi:10.1038/s41558-022-75001384-8.

Trout, T. J. 1996. Furrow irrigation erosion and sedimentation: on-field distribution. *Transactions of the American Society of Agricultural Engineers* 39, 1717-1723.

Uber, M., Haller, M., Brendel, C., Hillebrand, G. and Hoffmann, T. 2023. Past, present and future rainfall erosivity in Central Europe based on convection-permitting climate simulations. *Hydrology and Earth System Sciences Discussion* 2023, 1-30. doi:10.5194/hess-2023-120.

Vallebona, C., Pellegrino, E., Frumento, P. and Bonari, E. 2015. Temporal trends in extreme rainfall intensity and erosivity in the Mediterranean region: A case study in southern Tuscany, Italy. *Climatic Change* 128(1-2), 139-151.

van Vuuren, D. P., Edmonds, J., Kainuma, M., Riahi, K., Thomson, A., Hibbard, K., Hurtt, G. C., Kram, T., Krey, V., Lamarque, J.-F., Masui, T., Meinshausen, M., Nakicenovic, N., Smith, S. J. and Rose, S. K. 2011. The representative concentration pathways: An overview. *Climatic Change* 109(1-2), 5-31. doi:10.1007/s10584-011-0148-z.

Wallach, R., Ben-Arie, O. and Graber, E. R. 2005. Soil water repellency induced by long-term irrigation with treated sewage effluent. *Journal of Environmental Quality* 34(5), 1910-1920.

Wang, L., Zhang, F., Wang, G., Zeng, C., Chen, Y., Shi, X., Tang, H., Zhao, G., Xu, C. and Li, X. 2023. Response of soil erosion to climate and subsequent vegetation changes in a high-mountain basin. *Sustainability* 15(4), 3220.

Westerling, A. L., Hidalgo, H. G., Cayan, D. R. and Swetnam, T. W. 2006. Warming and earlier spring increase Western U.S. forest wildfire activity. *Science* 313(5789), 940-943.

Westra, S., Fowler, H. J., Evans, J. P., Alexander, L. V., Berg, P., Johnson, F., Kendon, E. J., Lenderink, G. and Roberts, N. M. 2014. Future changes to the intensity and frequency of short-duration extreme rainfall. *Reviews of Geophysics* 52(3), 522-555, doi:10.1002/2014RG000464.

Winterrath, T. 2022. Jährlicher (2001-2019) R-Faktor [N/h/yr] auf Basis der stündlichen Niederschlagszeitreihen der RADKLIM-Version 2017.002. Available at: https://opendata.dwd.de/climate_environment/CDC/grids_germany/annual/erosivity/precip_radklim/2017_002/ [last access: 15 May 2023].

Wischmeier, W. H. 1959. A rainfall erosion index for a universal soil-loss equation. *Soil Science Society of America Proceedings* 23(3), 246-249.

Wischmeier, W. H. 1960. Cropping-management factor evaluations for a Universal Soil-Loss Equation. *Soil Science Society of America Proceedings* 24(4), 322-326.

Wischmeier, W. H. and Smith, D. D. 1958. Rainfall energy and its relationship to soil loss. *Transactions of the American Geophysical Union* 39(2), 285-291.

Wischmeier, W. H. and Smith, D. D. 1978. Predicting rainfall erosion losses: A guide to conservation planning. *Agriculture Handbook* 537, United States Department of Agriculture, Washington, DC.

Xie, Y., Yin, S.-Q., Liu, B.-Y., Nearing, M. A. and Zhao, Y. 2016. Models for estimating daily rainfall erosivity in China. *Journal of Hydrology* 535, 547-558. doi:10.1016/j.jhydrol.2016.02.020.

Xu, X., Scanlon, B. R., Schilling, K. and Sun, A. 2013. Relative importance of climate and land surface changes on hydrologic changes in the US Midwest since the 1930s: implications for biofuel production. *Journal of Hydrology* 497, 110-120.

Xu, X., Yan, Y., Dai, Q., Yi, X., Hu, Z. and Cen, L. 2023. Spatial and temporal dynamics of rainfall erosivity in the karst region of southwest China: interannual and seasonal changes. *CATENA* 221, 106763.

Yoder, D. C., Porter, J. P., Laflen, J. M., Simanton, J. R., Renard, K. G., McCool, D. K. and Foster, G. R. 1997. *Cover-Management Factor (C). Predicting Soil Erosion by Water: A Guide to Conservation Planning with the Revised Universal Soil Loss Equation (RUSLE)*. US Department of Agriculture Handbook 703, 143-182.

Zuzel, J. F., Allmaras, R. R. and Greenwalt, R. 1982. Runoff and soil erosion on frozen soils in northeastern Oregon. *Journal of Soil and Water Conservation* 37, 351-354.

Part 2

Measuring soil erosion

Chapter 4

Advances in proximal instrumental techniques for measuring gully and rill erosion of agricultural soils

J. Casalí, M. A. Campo-Bescós, J. Álvarez-Mozos and R. Giménez, University of Navarre (UPNA), Spain*

1 Introduction

Ephemeral gullies and erosion rills are the two major types of concentrated flow erosion that occur on croplands. Ephemeral gully and rill erosion have significant impacts on agricultural land throughout the world. This chapter begins with a brief overview of the definitions, classification, and main characteristics associated with these types of erosion forms. The chapter then reviews characterization and quantification of these phenomena using either 3D or 2D approaches. 3D techniques are able to cover large areas but with lower measurement accuracy, while 2D techniques are more accurate but more labor intensive and can only be utilized on small areas.

Focusing on 2D techniques, the chapter then summarizes the current state of the art of ephemeral gully/rill measurement techniques based on the

* Institute for Innovation & Sustainable Development in Food Chain (IS-FOOD), Department of Engineering, Los Olivos building, Universidad Pública de Navarra, Campus de Arrosadia, 31006 Pamplona, Navarre, Spain.

http://dx.doi.org/10.19103/AS.2023.0131.06

characterization of channel cross-sections. Gully/rill erosion is characterized and channel volume measuring techniques are described. Factors contributing to measurement error are explored, including resolution of cross-section elevation profiles, channel size and shape, and number of cross-sections used per unit channel length. In addition, a case study is presented in which the real width of rills is obtained in field conditions for the first time. This shows that the authors' approach accurately determines the cross-section geometry of small erosion channels.

2 Ephemeral gully erosion

Gully erosion, historically defined as permanent gullies, is an important type of erosion. These gullies are caused by concentrated and intermittent flows of water that occur during and immediately following heavy rains and give rise to channels deep enough to interfere with normal agricultural tillage operations and to prevent their removal by such routine operations, making them a permanent feature on the landscape (Soil Science Society of America, 1984). Gullies affect both agricultural and non-agricultural environments. The classification of this type of water erosion is highly topical, as it affects the characterization and understanding of the diverse and complex processes involved in this type of erosion, and its subsequent prediction and modeling (Thwaites et al., 2022).

It is only relatively recently that so-called ephemeral gully erosion has been taken into consideration (Fig. 1). Currently, most authors (Foster, 1986; Poesen and Govers, 1990; Casalí et al., 1999; Foster, 2004) consider ephemeral gullies to be erosion channels formed by concentrated surface flow with the following characteristics:

1. Exclusively located on cultivated plots;
2. Periodically eliminated by conventional agricultural tillage, therefore having an ephemeral presence in the field;
3. Located in the swale that forms at the confluence of two opposing slopes (typically the valley bottom), tending to appear in the same positions year after year, in contrast to rills, which typically appear in random locations with varying sequences on slopes (consequently, the trajectory of the ephemeral gullies usually crosses that of the rill and the areas between rills) (Fig. 2);
4. Water flows through them only during heavy rain; and
5. Generally limited to a depth of no more than 25–30 cm by a tillage compaction layer in the lower stratum, and in many cases they are wide and shallow.

The use of the term 'ephemeral gullies' is relatively recent. In fact, Foster (1982) identified the need to give these particular, peculiar gullies, which at the time

Figure 1 Typical ephemeral gullies in cultivated areas of Navarre, Spain.

were called 'concentrated flow erosion' or 'erosion in waterways', a specific name. Once specified, ephemeral gully erosion, which frequently occurs all over the world, was identified as responsible for significant soil losses in agricultural areas and in many locations was the leading cause of soil loss and the most important source of sediment (USDA-NRCS, 1997; Vanmaercke et al., 2021).

Figure 2 Rills on a hillslope discharge to an ephemeral gully located in a swale in Bardenas Reales (Navarre, Spain).

Erosion by ephemeral gullies can cause significant damage to agricultural soils, both directly (through powerful erosive action) and indirectly (by facilitating the flushing of sediments generated by splash erosion and unconcentrated flow). This is because erosion channels increase connectivity in the movement of materials (Poesen and Hooke, 1997; Poesen et al., 2003), thus facilitating sediment transfer and loss.

The environmental impact of this phenomenon is not just limited to reduced productive capacity of the soil and the movement of the affected land but also to the distant impacts caused by the subsequent deposition of the eroded material. The economic losses caused by the clogging and/or pollution of lakes, rivers, and dams often exceed the severity of losses on the land. It is therefore not surprising that gully erosion is an issue of global importance (De Santisteban et al., 2006), and that erosion by ephemeral gullies in particular is considered the main problem that must be solved on agricultural land in North America (Alonso, 2007). Similarly, studies (e.g. Poesen and Valentin, 2003; Poesen and Hooke, 1997) show that gully erosion is an important agent of soil and environmental degradation in Europe (Poesen et al., 2006), especially in Mediterranean environments (Poesen et al., 2003).

As ephemeral gullies appear in the same location year after year, the drainage network becomes increasingly entrenched in the landscape. The continual downcutting of the gullies lowers the base level of slopes lateral to the gully, making them shorter and steeper, which can in turn increase the intensity of rill erosion and sheet erosion on these slopes (Foster, 1986). The severity of the problem is often masked by routine fills through tillage.

As mentioned earlier, ephemeral gully erosion was first studied and identified in the United States (Smith, 1993; Bandara and Sheshukov 2017; Daggupati et al., 2013; Daggupati et al., 2010; Daggupati et al., 2014; Islam et al., 2019; Karimov et al., 2014a,b; Foster, 1986; Grissinger and Murphey, 1989), but studies soon extended to other countries using different methods (Casalí et al., 2006, 2009; De Santisteban et al., 2006; Capra and la Spada 2002, 2012, 2013, 2015; Castillo and Gómez, 2016; Vanmaercke et al., 2021; Liu et al., 2021; Nachtergaele and Poesen, 1999; Nachtergaele et al., 2002; Poesen et al., 2003).

Despite the significant literature now available on the topic, there are still gaps in our understanding of this problem at a larger spatial scale. The available data refer to highly specific locations, typically over very short periods of time. In this regard, detailed field observations to evaluate the very few existing models (Douglas-Mankin et al., 2020) on the occurrence and evolution of ephemeral gully networks are also extremely scarce (Luquin et al., 2021). Due to the significant undertaking required to collect a long-term comprehensive dataset for gully erosion, public institutions are ideally positioned to take responsibility for continuous monitoring and data collection on representative areas.

It is unknown what effect mechanical tillage techniques, which became standard in the United States and Europe at the beginning of the twentieth century, have had on this type of erosion. Effective and efficient tillage may have resulted in more frequent exposure of bare soil to rainfall, increasing the risk of erosion. In addition, identifying and measuring gullies in an agricultural landscape has many challenges. Channels are relatively small and located amongst vegetation, making them hard to locate and measure, especially on a large scale and from a distance. There is also a temporal limitation for the study of ephemeral gullies because they are frequently eliminated by tillage. The time window for detecting and measuring ephemeral gullies is often very small, especially in the management of crops that involve frequent tillage, thus eliminating gullies. Conversely, studies indicate that conservation agriculture significantly reduces ephemeral gully erosion.

Based on field observations, three main sub-types of ephemeral gullies have been identified (Casalí et al., 1999):

1. Classical ephemeral gullies, considered the prototype ephemeral gully, are formed by concentrated runoff flows within the same field where runoff originated.
2. Drainage ephemeral gullies are created by concentrated flows from drainage areas above the field. Drainage flows reach the upstream end of a field and erode cultivated plots downstream. Channel incision begins at the very upper boundary of the plot.

3 Discontinuity ephemeral gullies typically form in places with an abrupt change in slope, such as field boundaries adjacent to roads with a small height-drop. This discontinuity contributes to the formation of gully headcuts.

More recently Chahor (2017) has highlighted the importance of ephemeral drainage gullies in an area of Navarre, Spain, representative of large areas in that region, indicating that prevalent ephemeral drainage gullies may be regionally significant.

Despite field and laboratory observations, many uncertainties remain regarding the processes and factors controlling the formation and function of ephemeral gullies (Bennett, 1999, Bennett et al., 2000a; Bennett and Casalí, 2001; Wells et al., 2009a,b; Wells et al., 2011). While studies show that ephemeral gully formation is often initiated by the development and migration of a headcut, other studies demonstrate that processes such as suffosion, seepage, and soil saturation also contribute to gully occurrence (Bernatek-Jakiel and Poesen, 2018).

3 Rill erosion

While ephemeral gully erosion has been recognized only recently, rill erosion was identified much earlier (Fig. 3). Occurring primarily on agricultural land, rills are arranged on sloping hillsides as a set of multiple, small (around 15 cm wide and 10 cm deep) (Foster, 1986), parallel and disconnected channels, more or less evenly distributed, flowing downhill and draining into small gullies,

Figure 3 Rill erosion on a hillslope in Bardenas Reales (Navarre, Spain). Some interrill areas are shown.

terraces, or depositional areas as part of so-called overland flow. Generally, individual rills do not appear in the same place. It is also common for rills to develop from tillage marks (Fig. 4).

Spaces between the rills are called interrill areas, and erosion within them is called interrill erosion (Fig. 3). According to the most accepted theory (Foster and Meyer, 1975), any erosion that occurs in these areas is known as rill erosion, so rill erosion and interrill erosion are grouped together. This approach is also adopted in the RUSLE model (Renard et al., 1997). This combined definition of rill and interrill erosion is widely accepted, although some authors refer to interrill erosion as laminar erosion. The effects of splashing within rills are generally neglected. In interrill areas, the impact of raindrops removes soil particles, which are then moved laterally by overland flow toward the rills

Figure 4 Rill erosion in a potato field around Leuven, Belgium. The effect of the rills, an impact of erosion due to tillage, can be seen.

(Foster, 1982), where most sediment movement takes place (Foster and Meyer, 1972). Since tillage in cropland areas removes the rills each year, rill erosion and interrill erosion remove soil uniformly at a given position on the slope. Due to its considerable erosive power, rill erosion can contribute significantly to the volume of sediment eroded from a slope, depending on the distance between rills and the extent of the area affected. Govers and Poesen (1988) found that in a cultivated area near Leuven, Belgium, material eroded in rills accounted for 54-78% of total erosion.

There are numerous laboratory studies dedicated to studying the processes and evolution of rill networks (Momm et al., 2018; Nearing et al., 1997; Giménez and Govers, 2001, 2004, 2007, Govers et al., 2007; Di Stefano et al., 2022), as well as models that deal specifically with the study of rills (Favis-Mortlock, 1998, 2000; Wu et al., 2018). However, the development of models for predictive purposes (i.e. not oriented to process research) has not progressed to date, since for decades rill estimation has been included in the USLE model and later RUSLE (Renard et al., 1997). Although there has been no pressing need to develop new models for rill erosion soil loss, ephemeral gully research lacks both process and predictive models.

It seems appropriate to distinguish between gullies and rills on the basis of the origin of the channels, either in swales or on slopes. Some authors have distinguished between rills and gullies according to whether the cross-sectional area exceeds a critical value. For example, Hauge (1977) establishes this critical value at 1 square foot (0.093 m^2). Others propose critical thresholds in terms of a minimum width and depth of 0.3 m and 0.6 m, respectively (Brice and Pecora, 1966) or a minimum depth of 0.5 m (Imeson and Kwaad, 1980). Piest (1980) warned of a lack of general agreement on dimensions that would distinguish a gully from a rill. In our opinion, it is more appropriate to consider the origin of the channels in order to classify them as gullies or rills. Rills are defined as channels formed on hillside slopes, and ephemeral gullies are those formed between hillside slopes, and their derivations, in watercourses. This approach is consistent with the random and variable nature with which rills are located on the same hillside from one rainfall event to another, since their formation will be subject to the high spatial variability of intrinsic soil factors (structural stability, hydraulic conductivity, etc.) as well as tillage. However, the location of an ephemeral gully on the landscape will depend primarily on topographic factors that direct the runoff trajectory, leading to the formation of this type of gully in the expected location.

In general, it is assumed that the hydraulic and erosion processes that occur in streams are similar in rills and ephemeral gullies (Bennett et al., 2000b). However, some studies suggest that there may be a functional difference between a rill and an ephemeral gully (Giménez and Govers, 2001; Giménez et al., 2004; Govers et al., 2007). In a rill, shallow flow causes a strong interaction between that stream and its channel, giving rise to typical stream macroforms, a succession of small concavities (pools) and straight sections (steps) (Giménez et al., 2019; Di Stefano et al., 2022) with a complex - and largely deterministic - spatial distribution along the channel (Morgan et al., in press). This interaction is ultimately responsible for why the hydraulic behavior of a rill differs from that observed in a river or a large channel. In a rill, e.g., the average flow velocity can be estimated independently of the slope grade (Govers, 1992; Takken et al., 1998; Giménez and Govers, 2001; Giménez et al., 2019; Di Stefano et al., 2018, 2019).

Would it therefore be possible to establish hydraulic rather than morphological criteria to separate rills from ephemeral gullies? The probable answer is yes. In fact, the typical step-pool morphology of rills and small erosion channels is increasingly recognized and studied, and specific hydraulic formulations for rills have been proposed (e.g. Di Stefano et al., 2022). It is worth noting that step-pool units, although of a different genesis to rills, are widely recognized and studied in certain mountain rivers (Chin and Wohl, 2005; Church and Zimmermann, 2007). Much of this knowledge could be applied to rill erosion.

4 Measurement of ephemeral gullies and rills

Although we can differentiate between ephemeral gullies and rills by their origin, many physical qualities can be considered together, such as an estimation of their volume and the temporal evolution of their channels, since they are similar in shape and size. Therefore, we will utilize this approach in this chapter. The precise measurement of their volume and channel evolution is essential for an adequate characterization of the process, its importance in the landscape, and to accurately and precisely determine erosion rates. In addition, detailed information on the shape and size of rill and gully features is necessary to develop and evaluate process-based simulation models.

Despite the fact that collecting physical measurements of feature size, shape, volume, and temporal evolution may seem relatively simple, new challenges arise, often in tandem with the new tools and methodologies available (Casalí et al., 2015; Castillo et al., 2019; Castillo et al., 2012). Two distinct overarching typologies exist among the available techniques for estimating the shape and volume of these channels. The first is those techniques based on in situ measurements of the shape of isolated cross-sections in channels (Casalí et al. 2006; De Santisteban et al. 2006; Sancho et al., 1991; Smith, 1993). Once the geometry of the section is determined, its area is calculated. The total volume of the studied channel is calculated using the expression:

$$V = \sum_{i=1}^{n} V_i = \sum_{i=1}^{n} \frac{A_{i-1} + A_i}{2} s \quad \text{Eq. 1}$$

where

V is the volume of eroded soil in the reach

n is the number of sub-reaches considered

V_i is the volume of eroded soil within the sub-reach i

A_{i-1} is the downstream cross-sectional area of the sub-reach

A_i is the upstream cross-sectional area of the sub-reach

s is the distance between adjacent cross-sections

These technologies are often referred to as 2D approximations.

Second are determinations made from point clouds and DEMs obtained in various ways, referred to as 3D approximations. These techniques can be further classified into those based on short distance observations and those based on long distance observations. Among the former group, techniques include photo-reconstruction, ground-based lidar, and total station techniques (Castillo et al., 2012). The latter group includes photogrammetry from aerial photographs (Giménez et al., 2009; D'Oleire-Oltmanns et al., 2012; Ries and Marzolff, 2003; Rock et al., 2012; Seeger et al., 2009) and satellite image analysis (Vanmaercke et al., 2021). Although they are included as a measurement technique, satellite images do not yet allow a sufficiently precise estimation of volume for the purposes proposed in this chapter.

We believe that the precise measurement of volume, shape, and evolution in the field by in-situ measurements from isolated cross-sections remains the most relevant method for the following reasons:

1 It is the most established method and still the most used. This also makes it possible to use old databases based on measurements using this method.
2 There are no feasible alternative methods when the soil is covered with vegetation. This is the case, e.g., in conservation tillage.
3 Lighting conditions often limit the ability to conduct measurements by image analysis. The accuracy of photogrammetric techniques is highly dependent on the time of day and/or time of year when the images are taken against channels with a width/depth ratio between 0.5 and 2.5, and the degree of error is unacceptable when this ratio is less than 0.5 (Giménez et al., 2009).
4 It is a reference method that can validate other methods. This was carried out, e.g., in Giménez et al. (2009).

There is a clear need for further evaluation of these methodologies.

The objective of this chapter is to review the current state of knowledge for the measurement of ephemeral gullies and rills via in-situ determinations from isolated cross-sections. In particular, this chapter is intended to provide guidance to those performing these measurements in practice. In addition, a case study is presented in which the real width of rills is obtained in field conditions for the first time.

5 Methods for determining the shape and size of ephemeral gully and rill cross-sections

This section briefly reviews the main methods available and traditionally used for the determination of the shape and size of the cross-sectional profile of ephemeral gully and rill sections. Special attention is given to the limitations and possibilities of each method.

The most basic method consists of a simple tape measure to survey the different lengths identified by the lead technician as representative of each cross-section. It has been used traditionally and is still used in certain circumstances (Casalí et al., 2006). Advantages of this method include:

- Its simplicity and robustness;
- Sufficient accuracy;
- Suitability in the presence of vegetation; and
- The ability to utilize it when the sections show hollows or cavities (see Fig. 5) in the slopes, making it useful as a complement to other methods that cannot make such estimates.

The main disadvantages include:

Figure 5 Two cross-sections facing downstream obtained from a non-erodible rill replica (slope: 8°; discharge: 5.5×10^{-4} m^3 s^{-1}). A large cavity is observed at the bottom of the cross-sections on the right. Source: Adapted from: Giménez et al. (2007).

Figure 6 Pin profilometers showing the cross-section of ephemeral gullies in Navarre, Spain. Distances *'L'* and *'d'* used in Equation 2 to calculate cross-sectional area are shown.

- The need to assume a simplified section geometry, which may result in estimates with significant errors if the section changes abruptly or is irregular;
- Manual recording and processing of data; and
- The need to draw up in advance an outline of each section, identifying the characteristic lengths that, in the technician's opinion, determine the section.

The next primary method is the pin profilometer (Sancho et al., 1991) (see Fig. 6), which consists of rods that are allowed to fall freely and vertically from a structure that contains and guides them, so that their upper ends represent the cross-sectional profile of the area to be measured. Once the ends of the rods have been arranged in this way, the panel can be photographed and the position of the top end of the rods can be determined in the frame with the aid of a graduated grid. With the help of a spreadsheet, the area of the complete section can be determined (e.g. Casalí et al., 1999; De Santisteban et al., 2006) using the following expression:

$$A = \sum_{i=1}^{n-1} A_i = \sum_{i=1}^{n} \frac{L_i + L_{i+1}}{2} d \quad \text{Eq. 2}$$

where

A is the cross-sectional area

n is the number of rods considered along a transverse section

Li is the distance from the top of the section to the bottom of the gully for rod *i*

d is the distance between adjacent rods

An even simpler approach is to place a sheet of paper on the flat bottom of the frame, mark the profile of the section, and determine the area of the section offsite (Smith, 1993).

The main advantages of this method are (Casalí et al., 2006):

- Its simplicity;
- Sufficient accuracy;
- Its applicability in the presence of vegetation; and
- The fact the profile can be photographed and obtained instantaneously, unlike more sophisticated methods such as the laser profilometer, studied below

Among the main disadvantages are:

- The need for manual data processing;
- The profilometer may be a heavy piece of equipment, making it difficult to transport within agricultural plots that are typically impossible to access with a vehicle;
- The need to level the device each time a section is measured; and
- The inability to record and measure excavated sections with cavities.

The final method is the profilometer equipped with a distancemeter (laser) (see Fig. 7). The set-up used by Giménez et al. (2009) consists of a 2.5 m long metal bar driven by computer-controlled stepping motors. In the field, the cross-section profile is determined at 5 mm intervals. The metal bar and the laser must be manually aligned and leveled over each of the selected cross-sections. The cross-sectional area can also be measured by applying Equation 2.

Advantages of this instrument include the following:

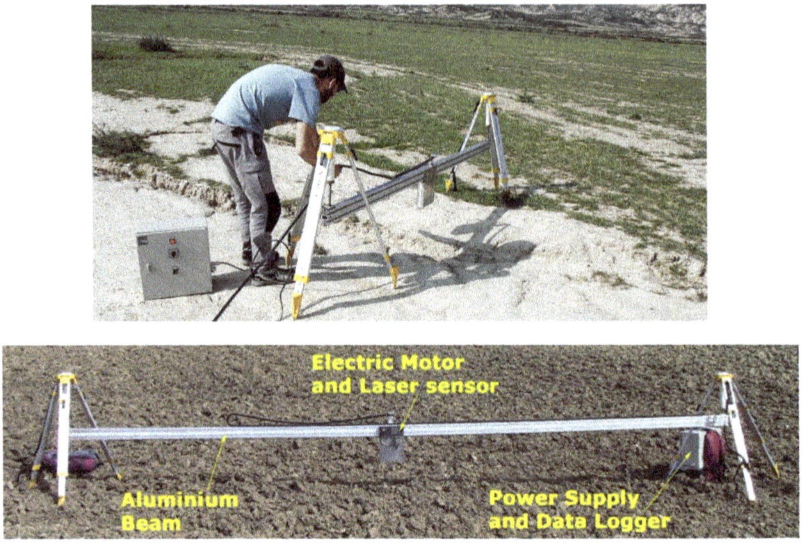

Figure 7 Laser profilometer measuring a cross-section (above). The main components of the device are shown (below).

- The profile can be highly detailed, since the distance between the measuring points can be very small (<5 mm); and
- The information in each profile is digitally recorded and therefore allows for automated processing.

Important disadvantages include:

- Greater complexity and subsequent potential for error, such as the lack of certain data due to beam deviation or loss of digital files;
- The significant time needed to position and level the device in each section prior to the measurement;
- The time needed for the laser sensor advance, which can be quite slow to process each section;
- Instrumentation can be heavy, bulky, and long, which makes it difficult to transport;
- Laboratory calibration and fine-tuning is required prior to field deployment;
- The technique is not directly applicable within vegetation; and
- It cannot record and measure excavated sections with cavities.

6 Issues in measuring shape and size

This section considers a range of issues which affect the accuracy of measurement:

- Effect of density of measurement points when estimating cross-sectional areas;
- Effect of the density of sections; and
- Determining the correct width of the gully/rill.

These are discussed in the following sections.

6.1 Effect of density of measurement points when estimating cross-sectional areas

The accuracy of the cross-sectional area measurement depends not only on the sensitivity of the measuring device but also on the level of detail of the corresponding elevation profile, i.e. on the density of measuring points or distance between two contiguous measurements. As we have seen, a classic, and still valid, technique for the field measurement of elevation cross-sections with high accuracy is the use of pin profilometers. However, due to limitations in their construction, the spacing between pins in these devices is 1-2 cm (Casalí et al., 2006). With modern laser profilometers capable of taking high resolution (i.e. several millimeters) readings available as a potential alternative, the degree of error for typical needle profilometers should be evaluated and assessed against the pros and cons of the laser profilometer listed earlier.

For this purpose, cross-sections of ephemeral gullies of contrasting size were selected from a database of an experimental study in which the authors of this chapter participated (Giménez et al., 2009). These sections were measured with a laser profilometer at measurement intervals of 5 mm and a height measurement accuracy (z-coordinate) of 0.5 mm (Giménez et al., 2009). Using these measurements, the areas of the cross-sections were calculated using Equation 2 and considered as reference values. Then, the cross-sectional elevation profile obtained was gradually decreased in resolution through the systematic elimination of measurement points, with a consequent increase in the equidistance between records (Fig. 8). Several cross-sectional elevation profiles were thus obtained for each of the selected sections, each with a different interval/equidistance between measurement points - from 10 mm to 200 mm, with a step of 5 mm. For each profile, the cross-sectional area was recalculated and the degree of error with respect to the reference value was determined (interval/equidistance = 5 mm,). Specifically, two different types of cross-sections were recognized and analyzed according to their cross-sectional area (A) and width-to-surface/depth ratio (WR): (1) small sections (A~0.25 mm^2, WR~0.3) and (2) large sections (A~0.45 m^2, WR~2.2).

In small sections (1) with measurement intervals of 10 mm, the degree of error was less than 2%, compared to intervals of 20 mm and 40 mm where the error was approximately 6% and 16%, respectively (Fig. 8, left column). In large sections (2), at intervals less than 40 mm, the error was less than 1%. As a

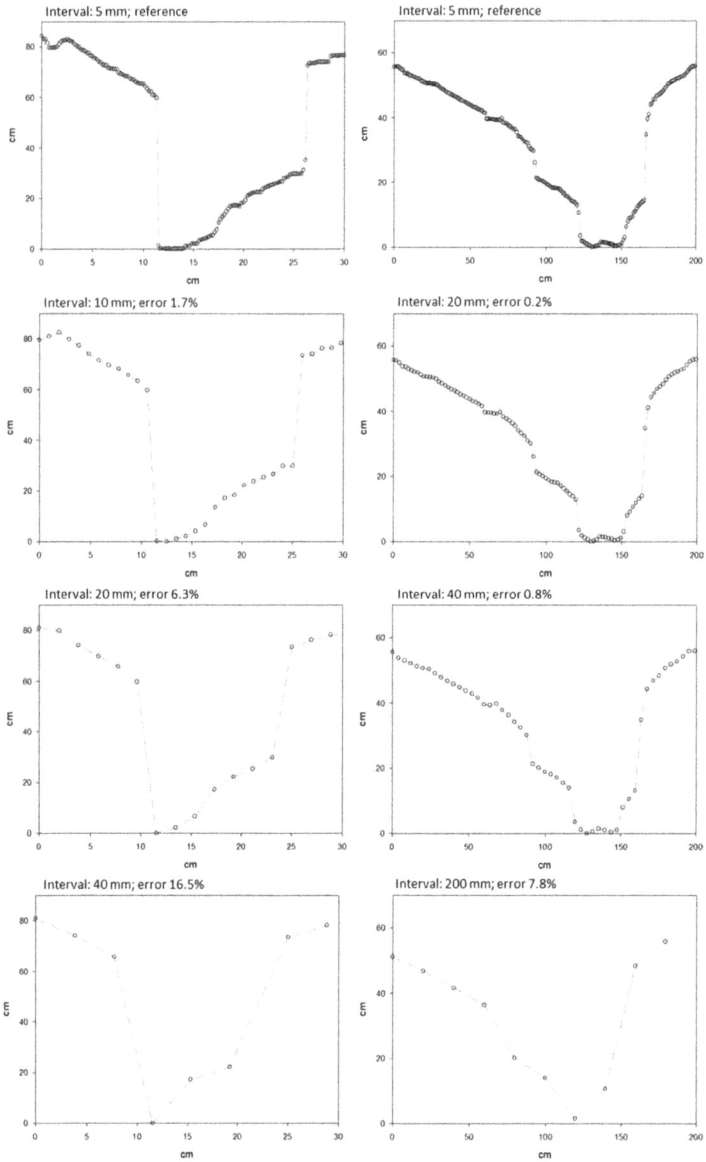

Figure 8 Cross-sections of two gullies obtained by using a laser profilometer (Source: Adapted from: Giménez et al., 2009) depicted at different resolutions (i.e. distances between measuring points). From each dataset, the cross-section area was measured and the percentage of variation (i.e. error) with respect to the reference (distance = 5 mm) was calculated. Left column: small gully (A~0.25 m^2, WR~0.3). Right column: large gully (A~0.45 m^2, WR~2.2). A: Cross-sectional area. WR: (upper) width/depth ratio. In each cross-section, both end-points of the imaginary line (not shown) that defines the section width were arbitrarily located at (almost) the top of each bank. Note the different scale in each figure.

maximum, errors in large sections approached 8% at a 200 mm interval (Fig. 8, right column).

These results suggest that the error in cross-sectional area determinations made with pin profilometers (pin spacing of ~10–20 mm, see above) is negligible in sections of no less than 0.5 m^2 and would be moderate in smaller sections. In the latter case, the degree of error would nevertheless be "acceptable' for the estimation of erosion rates (e.g. Casalí et al., 2006). For other purposes, such as comparative studies of field measurement methodologies (e.g. Giménez et al., 2009) where optimal reference values are required, the use of pin profilometers in small erosion channels could be questioned. Moreover, in large gullies (2), the use of tape measures for the measurement of cross-sections, as a preliminary step for the estimation of approximate values of erosion rates, would be acceptable (Casalí et al., 2006), tolerating the experimental errors inferred from the above (Fig. 8).

6.2 Effect of the density of sections

The 2D methods require the determination of the area of the cross-sections as well as the spacing (s) between them (Eq. 1); or, equivalently, the determination of the number of sections (in principle equidistant) to be measured in a given length of an ephemeral gully or rill. The question arises as to how many sections are necessary to obtain an accurate volume measurement. In other words, it is necessary to determine the appropriate density of observed cross-sections to obtain an acceptable error. This has already been explored by Casalí et al. (2006). Although this study requires completion and updating, it provides relevant information that can be summarized by stating that relative measurement errors above 10% were obtained very easily, even for small (s) values and for quasi-prismatic channels. Channel size and shape have a significant influence on measurement errors. Distance (s) must be <5 m to guarantee an error smaller than 10% even when the most precise cross-section area measurement method, a profilometer, is used.

6.3 Determining the correct width of the gully/rill

When measuring the area of a given cross-section by applying Equation 2, it is necessary to delimit the cross-section width first and then identify the remaining morphological parameters (e.g. depth and cross-section area). In many cases, the cross-section width is not obvious. For example, in Fig. 6, the horizontal line in blue, which determines the lengths Li, could be placed higher or lower without there being any clear criterion for this in principle. Needless to say, the value of the cross-sectional area is different depending on the position chosen (see Casalí et al., 2015 for a detailed reflection on this). This subjectivity when identifying the width also arises when sections are measured by tape measure. It is important to highlight that this problem also affects 3D methods for volume estimation based on point clouds, since it is always necessary to determine the limits of the gully.

Several methods have been proposed to deal with this problem by automating the process and making it more objective (Castillo et al., 2014, 2019; Palmeri et al., 2021). Because the criteria considered in these measurements are not clearly defined, the disparity is not apparent and the implicit error embedded in comparative studies is not obvious. This subject is extensively reviewed and discussed herein.

As has been pointed out in several sections of this chapter, the morphological characterization of erosion channels is essential to quantify soil losses and to analyze their dynamics. The main issue of this morphological characterization is associated with obtaining exact width and depth values for the eroded channel. The width is usually defined as the imaginary line whose ends are located in two abrupt changes of slope at both sides of the cross-section of the channel (Casalí et al., 2015). This criterion is applied to direct field measurements as well as when digital elevation models (DEMs) and ad hoc mathematic algorithms (e.g. Evans and Lindsay, 2010; Parker et al., 2012; Castillo et al., 2014) are employed in the characterization. The location of these ends is not always clear or evident, especially in cross-sections with irregular shapes and poorly defined limits.

The same problem occurs with depth, generally defined as the vertical distance between the lowest point of the profile and the imaginary line representing the width. However, in irregular bed cross-sections, assigning the minimum elevation point as representative of the entire section is questionable. Thus, the characterization of the cross-section of an eroded channel, especially the determination of its width, is conditioned by the particular perception of the expert. This results in significant problems in repeatability of measurements and leads to uncertainties when comparing experimental results of concentrated flow erosion obtained by different experts (Casalí et al., 2015). Significant differences can be obtained when calculating the volume of detached soil and, consequently, in erosion rate values.

Nevertheless, even in the case of a channel with an apparently clearly defined width, a question arises. Do the limits of the channel, defined as previously explained, really correspond to the transversal limits of the erosive process that caused the gully (Casalí et al., 2015)?

The only objective and unambiguous way to determine the morphology of an eroded channel is by superposition, at any location along the channel, of the elevation profiles obtained before and after the channel was formed (Casalí et al., 2015). The width of the section is determined by the two intersection points of the profiles, and the cross-section is the area defined by the intersection. A 3D characterization is obtained by repeating this operation throughout the channel. However, topographic information obtained prior to the erosive process taking place is rarely available.

7 Case study: *in situ* characterization of small, eroded channels

7.1 Introduction

This case study has allowed, for the first time, the cross-section geometry of small erosion channels (rills) to be explicitly determined (following Casalí et al., 2015). The characterization was entrusted to a group of experts in order to illustrate the common discrepancies and trends that occur between researchers and carried out using an automated approach recently proposed by Castillo et al. (2019). The findings presented herein can be directly extrapolated to larger erosion channels such as ephemeral gullies.

7.2 Material and methods

Detailed measurements of the geometry of small erosion channels (rills) were conducted from an experimental dataset previously obtained by some of the authors of this chapter (Giménez et al., 2019). A collection of six rills was created in the field from different combinations of slope (*S*) and discharge (*Q*): S = 5% and 15%; Q = 5 $m^3\ h^{-1}$, 2.5 $m^3\ h^{-1}$, 0.16 $m^3\ h^{-1}$, 0.36 $m^3\ h^{-1}$, 0.78 $m^3\ h^{-1}$, and 0.9 $m^3\ h^{-1}$. Several hundred overlapping photographs were taken from different angles covering the entire experimental area. A high-resolution DEM was created from these photographs using professional software (Agisoft® PhotoScan Professional edition v1.0, 2013) (Giménez et al., 2019). Twenty varying cross-sections were extracted from the DEMs, and DEMs were also available from before the rills were formed, covering the entire experimental area (Giménez et al., 2019).

For each selected cross-section, width was determined by superposition of the elevation profiles before and after erosion processes occurred (Fig. 9). The width was then defined as the imaginary line connecting the intersection points of both profiles, and the width length was defined as the horizontal projection of the imaginary line. The cross-section area was delimited by the elevation profiles and the imaginary line of the width (Fig. 9). Maximum depth was defined as the vertical distance from the imaginary line to the lowest elevation point. These geometrical parameters were considered as reference values and compared with those obtained by 11 independent experts.

The experts only had access to the cross-sectional elevation profiles obtained from DEMs following rill formation. For each cross-section, they determined the width as the imaginary line whose ends were located in abrupt changes of

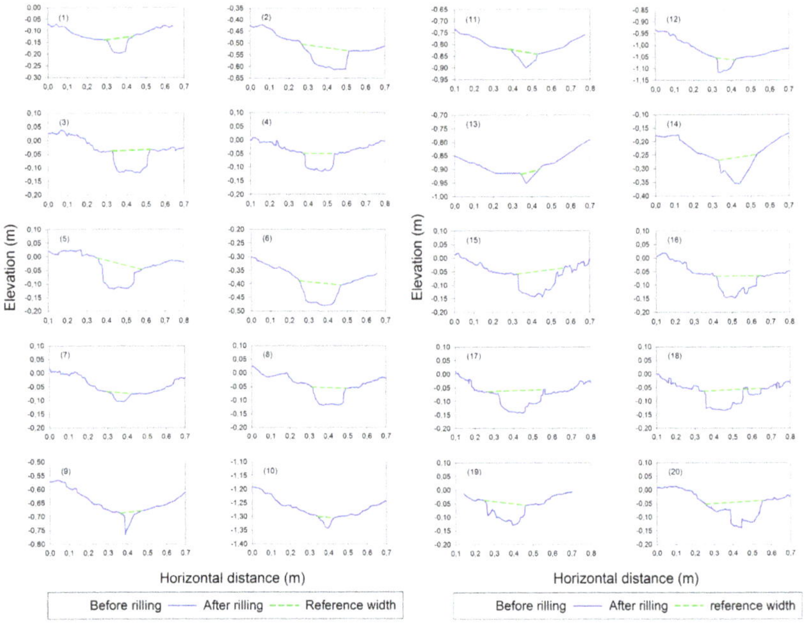

Figure 9 Elevation profiles before and after rill erosion for each of the 20 cross-sections. The reference width is the imaginary line (green dashed line) connecting the intersection points of both profiles. Elevation profiles before and after rill erosion for each of the 20 cross-sections. The reference width is the imaginary line (green dashed line) connecting the intersection points of both profiles.

slope at both sides of the cross-section, based on their own judgment. Width length, cross-sectional area, and maximum depth were defined as before. The experts were informed that each rill was the result of a single erosive pulse, but no data were provided on the combination of discharge and slope. The cross-sections were classified as (1) triangular, (2) rectangular, and (3) irregular, i.e. it did not fit with any regular geometrical shape. The experts did not have access to this classification. A descriptive statistical analysis of the results was then carried out.

For comparison purposes, the selected cross-sections were also analyzed using the Optimal Lid Method (Olid), an automated, non-threshold-dependent approach proposed by Castillo et al. (2019). This works by maximizing the sum of three different topographic parameters - prominent elevation, good margin alignment, and smooth pre-erosion surface (Castillo et al., 2019). It should be noted that the Olid method assumes by default that a gully cross-section is the result of multiple erosion events leading eventually to a succession of different 'terraces' or levels of erosion in the eroded channel. When this occurs, the uppermost boundary of the cross-section is that which is closest to the surface, and a nested cross-section, involving all the terraces, is then depicted.

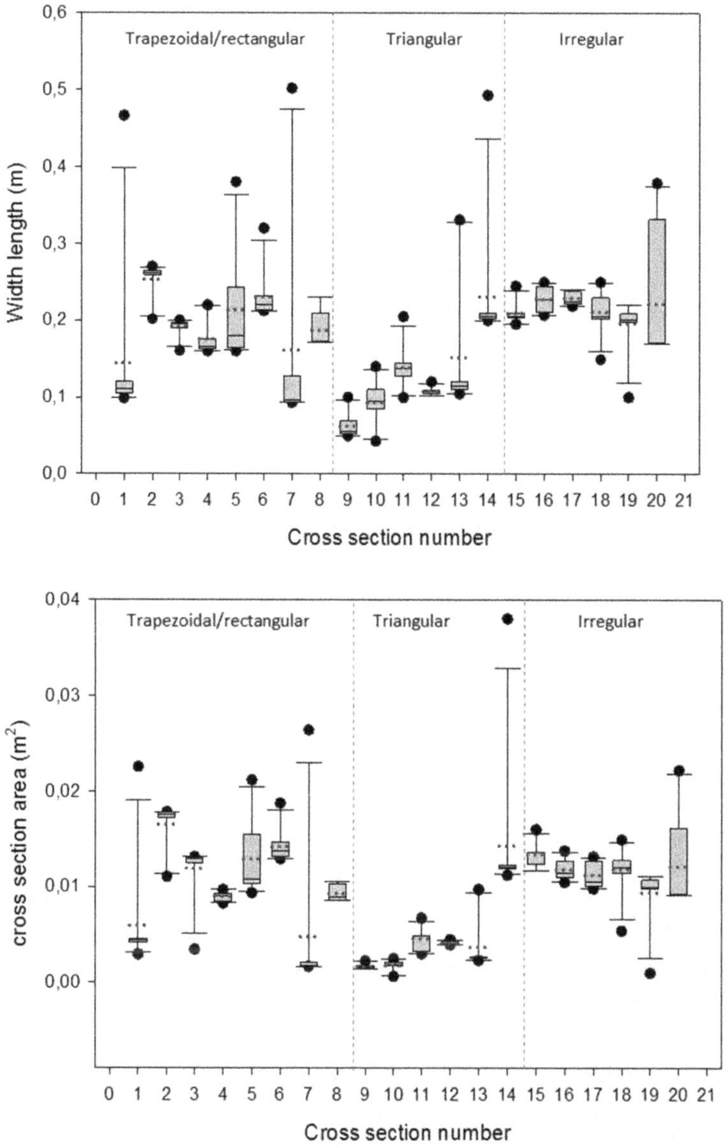

Figure 10 Width length (above) and cross-section area (below) measurements carried out by the experts in each of the cross-sections. Rectangular/trapezoidal, triangular, and irregular are the general shapes of the cross-sections. Black and red dotted lines inside boxes are median an average value, respectively. The black dots correspond to the highest and lowest value assigned by the experts in each section.

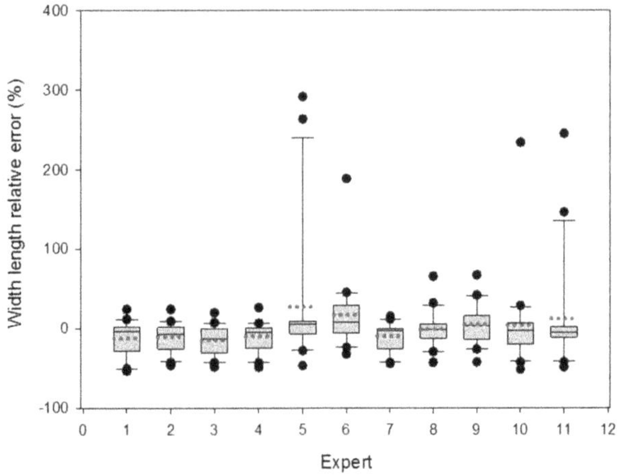

Figure 11 Relative error by the different experts when measuring channel width length. The relative error is the percentage difference between each of the 20 width length determinations with respect to the corresponding reference value. Black and red dotted lines inside boxes are median and average values, respectively.

7.3 Results and discussion: assessment of the experts' performance

The width length, depth, and area of the selected cross-sections, determined by the superposition of the DEMs, i.e. reference values (Fig. 9) were on average 0.18 m (0.065), 0.07 m (0.018), and 0.009 m^2 (0.005), respectively (standard deviation in brackets).

A disparity of criteria was observed among the experts when delimiting the cross-section widths (coefficient of variation, CV = 25%) and, consequently, its area (CV = 33%) (Figs 10 and 11). Variability did not appear to be associated with the general shape of the cross-section (Fig. 11). Castillo et al. (2019) also evaluated the variability among a group of experts when delimiting the geometry of different gullies – the mean CV for the width and cross-section area was around 17% and 12%, respectively.

Several reasons may explain the variability of criteria among experts. Channels with more uniform sloping walls and without abrupt changes in inclination hinder the identification of endpoints that delimit the width of the channel, resulting in a subjective and arbitrary task (e.g. Fig. 9, sections #10 and #13). In contrast, several abrupt changes in slope can be found in some irregular cross-sections (e.g. Fig. 9, sections #19 and #20) and choosing the right ones as the channel limit is also highly subjective (Castillo et al., 2019).

The experts understood that a single erosive event had formed each rill and therefore a succession of bank-plateau morphologies, i.e. terraces, was not expected. However, they had no information on the formative slope-discharge and, consequently, on the magnitude of the erosive pulse. This lack of 'perspective', or reference, on the erosive power of the flow hindered speculations on the expected size of the cross-sectional area. For example, a highly erosive flow precedes the formation of a relatively large cross-section area. Thus, the imaginary line that delimits the width of the channel would be located at a relatively high level of the elevation profile, subsequently increasing the maximum depth assumed for the eroded channel.

The impact of the variability in the expert criteria can be illustrated by estimating the volume of a fictitious 10 m-long eroded channel from some of the experimental cross-section areas determined by experts with contrasting points of view. For instance, considering cross-section #10 (Fig. 9), the corresponding volume would go from 0.010 m^3 to 0.16 m^3, respectively. Similarly, for cross-section #5 (Fig. 9), the minimum and maximum volume would be 0.015 m^3 and 0.43 m^3, respectively. Therefore, this would lead in turn to differences in erosion rates of an order of magnitude depending only upon the arbitrary criteria of the eventual researcher in charge of a fictitious survey.

7.4 Results and discussion: assessment of the performance of the Olid method

In general, the Olid method largely overestimated the width length, depth, and area of the cross-sections by 87%, 26%, and 95%, respectively (median values). However, the overestimation was mainly on those cross-sections presenting terraced banks, when the upper limit was considered closest to the surface since multiple erosion events were assumed to have occurred (see above) (Castillo et al., 2019). For instance, the width length and cross-section area of cross-section #17 (Fig. 10) were overestimated by around 235% and 300%, respectively, while on cross-section #14, with more regular banks (Fig. 10), both the width and cross-section area were only overestimated by less than 10%.

7.5 Results and discussion: insights for field geometry characterization of eroded channels

If this characterization was developed directly in-situ, in front of a rill or an ephemeral gully, the delimitation of the cross-section could be to some extent supported by the observation of the 'scars' of the soil detachment. When the eroded channel is the result of multiple erosion events, the channel width is normally located in the upper terrace, closest to the surface. However, multiple erosion events do not always result in terraced-like channels (see below). If it is known or suspected in advance that the eroded channel was formed primarily by the action of a single erosion event, then the cross-section limits should be located in the lowest terrace.

Generally, in irregular cross-sections, the line that defines the width of the channel is not horizontal, so both ends have to be separately defined on each margin. To more clearly define the upper limit of an irregular cross-section, the general shape of the landscape on both sides of the channel (at a distance no less than five-times the bottom width) should be taken into account.

The findings presented herein can be directly extrapolated to ephemeral gullies. However, because gullies are usually larger channels than rills, they are more susceptible to changes in a portion of the cross-section, due to, e.g., localized wall collapse. Since the cross-sections of gullies are relatively deep, several soil layers can be affected, each presenting a different mechanical response to the erosive agent. For instance, the cross-sections of cohesive soils are relatively narrow. This could generate irregular sections and eventually terraces, i.e. nested cross-sections that deceptively appear to be the result of multiple erosion events.

7.6 Conclusion

Significant subjectivity discrepancies were found among the experts in terms of how the cross-section geometry was estimated, which could lead to important differences in the estimation of erosion rates. Clear criteria to define and determine critical morphological characteristics of small erosion channels are still required despite the fact that these criteria are typically taken for granted. Mathematical tools are necessary to avoid, at a minimum, the human-related subjectivity that occurs when characterizing the geometry of gullies and rills. Ideally, these should be process-based approaches distinguishing, for instance, between single or multiple erosion events. To this end, further investigations are required for an improved, non-arbitrary definition of topographic parameters and their threshold values, strengthening current approaches.

8 Conclusion

Agricultural activity tends to increase soil water erosion, which also affects erosion by rills and ephemeral gullies. Erosion by ephemeral gullies or rills (via concentrated flow) can cause significant damage to agricultural soils, both directly (due to its powerful erosive action) and indirectly (by facilitating the movement of sediment generated by splash erosion and non-concentrated flow). The environmental impact of this type of phenomenon is not only limited to the reduced productive capacity of the soil and the movement of the affected land but also to the off-site effect caused by the subsequent deposition of eroded material. Indeed, the economic damage caused by the siltation and/or pollution of lakes, rivers and dams often exceeds the damage caused to the land. Therefore, it is not surprising that erosion by ephemeral gullies or rills is of global concern, being an important agent of soil and environmental degradation.

In order to study concentrated flow erosion through the measurement of the volume of gullies/rills, the correct characterization of the geometry of eroded channels is essential. An accurate and convenient way to conduct this is through in-situ measurements of representative cross-sections of the channels (2D approach). Accuracy in measuring the cross-sectional area will depend on the level of detail of its cross-sectional elevation profile; that is, on the density of measurement points. For example, and according to our data, accepting a 5% experimental error as valid, the distance between two contiguous measurements should not exceed 2 cm in cross-sections smaller than 0.25 m^2, while in larger sections ($\geq$~0.5 m^2) it should not exceed 20 cm. In addition, the presence of cavities in the channel wall should be estimated and weighed, at least approximately, since they will not be reflected in the cross-sectional elevation profile if it is developed from orthogonal measurements with respect to the channel base, i.e. when profilometers are used.

The definition of the area of a cross-section will depend not only on the accuracy of the elevation profile but also on the correct demarcation of the upper limit (width) of the cross-section. The delimitation of the width can be ambiguous, especially in sections of complex shapes, and its determination is therefore subject to the particular criterion of the researcher. However, it is surprising that researchers appear to be largely unaware of this lack of unifying criteria and thus of the error implicit in comparative studies. The only explicit way to make these determinations would be to have a cross-sectional elevation profile of the terrain prior to the erosion process. The superimposition of this with the elevation profile of the eroded cross-section would allow one to define the width using the points of intersection of the profiles. However, topographic information prior to the formation of the erosion channel is rarely available. Thus, it is imperative to define clear morphological criteria for the correct characterization of the geometry of erosion channels. It should be noted that, although the current mathematical algorithms are based on arbitrary morphological thresholds and therefore could be improved, they ensure, at least, objectivity in the records.

Previous studies by our research group determined that the distance between consecutive cross-sections should not exceed 5 m to ensure an error of less than ~10% in the estimation of channel volume.

In short, optimal measurement, in terms of precision and objectivity, of the volume of small erosion channels is not to be trivialized or restricted to the academic field. For example, two experts could reach significantly different conclusions in their estimations of erosion rates per ephemeral gully/rill on the same site, simply because they had different criteria when characterizing the cross-sections of the erosion channels. As a result, the land use and management strategies thus implemented, based on erosion estimates, would be equally different.

As a final consideration, we should not assume that estimates of concentrated flow erosion rates in the global literature are fully comparable with each other due to the channel estimation discrepancies specified in this chapter.

9 Future trends in research

Future research should investigate how to improve the parameterization of current erosion models with the high spatio-temporal resolution data provided by modern techniques for monitoring the erosion process. Further Reading: Epple et al. (2022).

In line with the previous paragraph, the development of new algorithms that delineate cross-sections of erosion channels objectively and accurately should be prioritized. Further Reading: Castillo et al. (2019).

10 Where to look for further information

- Castillo, C., Campo-Bescós, M. A., Giménez, R., Pérez, R. and Casalí, J. 2019. The Optimal Lid Method for the objective definition of cross-section limits in ephemeral gullies. *Catena* 176, 381–393.
- Epple, L., Kaiser, A., Schindewolf, M., Bienert, A., Lenz, J. and Eltner, A. 2022. A review on the possibilities and challenges of today's soil and soil surface assessment techniques in the context of process-based soil erosion models. *Remote Sensing* 2022, 14, 2468.

11 References

Alonso, C. V. 2007. Ephemeral gully erosion research: Processes and modeling. In: Casalí, J. and Giménez, R. (Eds.), *Progress in Gully Erosion Research*. Pamplona: Universidad Pública de Navarra. Nafarroako Unibertsitate Publikoa, pp. 15–17.

Bandara, C. and Sheshukov, A. Y. 2017. Quantifying ephemeral gully erosion with photogrammetry surveying. In: 2017 ASABE Annual International Meeting. American Society of Agricultural and Biological Engineers, Spokane, Washington. https://doi.org/10.13031/AIM.201700474.

Bennett, S. J. 1999. Effect of slope on the growth and migration of headcuts in rills. *Geomorphology* 30(3), 273-290. https://doi.org/10.1016/S0169-555X(99)00035-5.

Bennett, S. J. and Casalí, J. 2001. Effect of initial step height on headcut development in upland concentrated flows. *Water Resources Research* 37(5), 1475–1484. https://doi.org/10.1029/2000WR900373.

Bennett, S. J., Alonso, C. V., Prasad, S. N. and Römkens, M. J. M. 2000a. Experiments on headcut growth and migration in concentrated flows typical of upland areas. *Water Resources Research* 36(7), 1911–1922. https://doi.org/10.1029/2000WR900067.

Bennett, S. J., Casalí, J., Robinson, K. M. and Kadavy, K. C. 2000b. Characteristics of actively eroding ephemeral gullies in an experimental channel. *Transactions of the American Society of for Agricultural Engineers* 43, 641-649. https://doi.org/10.13031/2013.2745.

Bernatek-Jakiel, A. and Poesen, J. 2018. Subsurface erosion by soil piping: Significance and research needs. *Earth-Science Reviews* 185, 1107-1128. https://doi.org/10.1016/J.EARSCIREV.2018.08.006.

Brice, J. C. and Pecora, W. T. 1966. Erosion and deposition in the loess-mantled Great Plains, medicine creek drainage basin, Nebraska, U.S. *Professional Paper* 352H, 235-339. https://doi.org/10.3133/pp352H.

Capra, A. and la Spada, C. 2015. Medium-term evolution of some ephemeral gullies in Sicily (Italy). *Soil and Tillage Research* 154, 34-43. https://doi.org/10.1016/J.STILL.2015.07.001.

Capra, A. and Scicolone, B. 2002. SW—soil and water: Ephemeral Gully Erosion in a Wheat-cultivated Area in Sicily (Italy). *Biosystems Engineering* 83(1), 119-126. https://doi.org/10.1006/BIOE.2002.0092.

Capra, A., Ferro, V., Porto, P. and Scicolone, B. 2012. Quantifying interrill and ephemeral gully erosion in a small Sicilian basin. *Zeitschrift fur Geomorphologie, Supplementary Issues* 56(1), 9-25. https://doi.org/10.1127/0372-8854/2012/S-00070.

Casalí, J., Giménez, R. and Bennett, S. 2009. Gully erosion processes: Monitoring and modelling. *Earth Surface Processes and Landforms* 34(14), 1839-1840.

Casalí, J., Giménez, R. and Campo-Bescós,M. A. 2015. Gully geometry: What are we measuring? *SOIL* 1(2), 509-513.

Casalí, J., Loizu, J., Campo, M. A., de Santisteban, L. M. and Álvarez-Mozos, J. 2006. Accuracy of methods for field assessment of rill and ephemeral gully erosion. *Catena* 67(2), 128-138. https://doi.org/10.1016/J.CATENA.2006.03.005.

Casalí, J., López, J. J. and Giráldez, J. V. 1999a. Ephemeral gully erosion in southern Navarra (Spain). *Catena* 36(1-2), 65-84. https://doi.org/10.1016/S0341-8162(99)00013-2.

Casalí, J., López, J. J. and Giráldez, J. V. 1999b. Ephemeral gully erosion in southern Navarra (Spain). *Catena* 36(1-2), 65-84. https://doi.org/10.1016/S0341-8162(99)00013-2.

Castillo, C. and Gómez, J. A. 2016. A century of gully erosion research: Urgency, complexity and study approaches. *Earth-Science Reviews* 160, 300-319. https://doi.org/10.1016/J.EARSCIREV.2016.07.009.

Castillo, C., Taguas, E. V., Zarco-Tejada, P., James, M. R. and Gómez, J. A. 2014. The normalized topographic method: An automated procedure for gully mapping using GIS. *Earth Surface Processes and Landforms* 39(15), 2002-2015. https://doi.org/10.1002/ESP.3595.

Castillo, C., Campo-Bescós, M. A., Giménez, R., Pérez, R. and Casalí, J. 2019. The Optimal Lid Method for the objective definition of cross-section limits in ephemeral gullies. *Catena* 176, 381-393. https://doi.org/10.1016/J.CATENA.2019.01.032.

Chahor, Y. 2017. *Analysis of the Hydrological and Erosive Behavior of Agrarian Watersheds in Navarre (Spain) by Experimentation and Modelling*. Pamplona: Public University of Navarre.

Chin, A. and Wohl, E. 2005. Toward a theory for step pools in stream channels. *Progress in Physical Geography: Earth and Environment* 29(3), 275-296. https://doi.org/10.1191/0309133305PP449RA.

Church, M. and Zimmermann, A. 2007. Form and stability of step-pool channels: Research progress. *Water Resources Research* 43(3). https://doi.org/10.1029/2006WR005037.

Daggupati, P., Douglas-Mankin, K. R., Sheshukov, A. Y. and Barnes, P. L. 2010. Monitoring and estimating ephemeral gully erosion using field measurements and GIS. In: American Society of Agricultural and Biological Engineers Annual International Meeting 2010, pp. 5318–5332. https://doi.org/10.13031/2013.29976.

Daggupati, P., Douglas-Mankin, K. R. and Sheshukov, A. Y. 2013. Predicting ephemeral gully location and length using topographic index models. *Transactions of the ASABE* 56, 1427–1440. https://doi.org/10.13031/TRANS.56.10087.

Daggupati, P., Sheshukov, A. Y. and Douglas-Mankin, K. R. 2014. Evaluating ephemeral gullies with a process-based topographic index model. *Catena* 113, 177–186. https://doi.org/10.1016/J.CATENA.2013.10.005.

de Santisteban, L. M., Casalí, J. and López, J. J. 2006. Assessing soil erosion rates in cultivated areas of Navarre (Spain). *Earth Surface Processes and Landforms* 31(4), 487-506. https://doi.org/10.1002/ESP.1281.

di Stefano, C., Nicosia, A., Palmeri, V., Pampalone, V. and Ferro, V. 2022. Rill flow resistance law under sediment transport. *Journal of Soils and Sediments* 22(1), 334-347. https://doi.org/10.1007/S11368-021-03083-X.

Di Stefano,C., Ferro, V., Palmeri, V. and Pampalone,V. 2018. Testing slope effect on flow resistance equation for mobile bed rills. *Hydrological Processes* 32(5), 664–671.

Di Stefano,C., Palmeri, V., Pampalone,V. and Ferro, V. 2019. Dissipative analogies of step-pool features: From rills to mountain streams. *Catena* 174, 235–247.

D'Oleire-Oltmanns, S., Marzolff, I., Peter, K. D. and Ries, J. B. 2012. Unmanned aerial vehicle (UAV) for monitoring soil erosion in morocco. *Remote Sensing* 4(11), 3390–3416. https://doi.org/10.3390/RS4113390.

Douglas-Mankin, K. R., Roy, S. K., Sheshukov, A. Y., Biswas, A., Gharabaghi, B., Binns, A., Rudra, R., Shrestha, N. K. and Daggupati, P. 2020. A comprehensive review of ephemeral gully erosion models. *Catena* 195. https://doi.org/10.1016/J.CATENA.2020.104901.

Evans, M. and Lindsay, J. 2010. High resolution quantification of gully erosion in upland peatlands at the landscape scale. *Earth Surface Processes and Landforms* 35(8), 876–886. https://doi.org/10.1002/ESP.1918.

Favis-Mortlock,D. T. 1998. *A Self-Organizing Dynamic Systems Approach to Hillslope Rill Initiation and Growth: Model Development and Validation*. IAHS[MS1] Publication. Series of Proceedings and Reports (Red Book Series) Volume: 249 (1998) ISSN: 0144-7815 Online ISSN: 2731-2887.

Favis-Mortlock,D. T., Boardman, J., Parsons, A. J. and Lascelles,B. 2000. Emergence and erosion: A model for rill initiation and development. *Hydrological Processes* 14(11–12), 2173–2205.

Foster, G. R. 2004. Modeling ephemeral gully erosion for conservation planning. In: Proceedings of the III International Symposium. Oxford, Mississippi, pp. 118–149.

Foster, G. R. 1982. Channel erosion within farm fields, Preprints – ASCE Convention & Exposition. New York: ASCE.

Foster, G. R. 1986. Understanding ephemeral gully erosion. In: National Academies Press (Ed.), *Soil Conservation*. Washington, DC, pp. 90–125. https://doi.org/10.17226/648.

Foster, G. R. and Meyer, L. D. 1972. Transport of soil particles by shallow flow. *Transactions of the American Society of for Agricultural Engineers* 15, 99–102. https://doi.org/10.13031/2013.37840.

Foster, G. R. and Meyer, L. D. 1975. Mathematical simulation of upland erosion by fundamental erosion mechanics. In: Present and Prospective Technology for Predict Sediment Yields and Sources. US Dep of Agric, Agric Res Serv (ARS-S-40), Washington, DC, pp. 190–207.

Giménez, R. and Govers, G. 2001. Interaction between bed roughness and flow hydraulics in eroding rills. *Water Resources Research* 37(3), 791–799. https://doi.org/10.1029/2000WR900252.

Giménez, R., Planchon, O., Silvera, N. and Govers, G. 2004. Longitudinal velocity patterns and bed morphology interaction in a rill. *Earth Surface Processes and Landforms* 29(1), 105–114. https://doi.org/10.1002/ESP.1021.

Giménez, R., Léonard, J., Duval, Y., Richard, G. and Govers, G. 2007. Effect of bed topography on soil aggregates transport by rill flow. *Earth Surface Processes and Landforms* 32(4), 602–611. https://doi.org/10.1002/ESP.1418.

Giménez, R., Marzolff, I., Campo, M. A., Seeger, M., Ries, J. B., Casalí, J. and Álvarez-Mozos, J. 2009. Accuracy of high-resolution photogrammetric measurements of gullies with contrasting morphology. *Earth Surface Processes and Landforms* 34(14), 1915–1926. https://doi.org/10.1002/ESP.1868.

Giménez, R., Zubieta, E., Govers, G. and Campo-Bescos, M. A. 2019. In-situ assessment of the spatial arrangement of step-pool units on eroded rills. *Earth Surface Processes and Landforms* 44(14), 2731-2740. https://doi.org/10.1002/ESP.4700.

Govers, G. 1992. Relationship between discharge, velocity and flow area for rills eroding loose, non-layered materials. *Earth Surface Processes and Landforms* 17, 515–528.

Govers, G. and Poesen, J. 1988. Assessment of the interrill and rill contributions to total soil loss from an upland field plot. *Geomorphology* 1(4), 343–354. https://doi.org/10.1016/0169-555X(88)90006-2.

Govers, G., Giménez Díaz, R. and van Oost, K. 2007. Implications of recent experimental findings for rill erosion modeling. In: Casalí, J. and Giménez, R. (Eds.), *Progress in Gully Erosion Research*. Pamplona: Universidad Pública de Navarra. Nafarroako Unibertsitate Publikoa, pp. 56–57.

Grissinger, E. and Murphey, J. 1989. Ephemeral gully erosion in the loess uplands, Goodwin Creek watershed, northern Mississippi, USA. In: China Ocean (Ed.), International Symposium on River Sedimentation. 4. Beijing, pp. 51–58.

Hauge, C. 1977. Soil erosion definitions. *California Geology* 30, 202–203.

Imeson, A. C. and Kwaad, F. J. P. M. 1980. Gully types and gully prediction. *Geografisch Tijdschrift* 14, 430–441.

Islam, M. A., Weerasekara, C., Zhou, Y. and Sheshukov, A. Y. 2019. Ephemeral gully erosion modeling and surveying with close-range digital photogrammetry. In: 2019 ASABE Annual International Meeting. American Society of Agricultural and Biological Engineers. https://doi.org/10.13031/AIM.201901426.

Karimov, V., Sheshukov, A. and Barnes, P. 2014. In situ field study of ephemeral gully propagation in a cultivated cropland. In: American Society of Agricultural and Biological Engineers Annual International Meeting 2014, ASABE 2014, pp. 4238-4244. https://doi.org/10.13031/aim.20141912018.

Karimov, V., Sheshukov, A. and Barnes, P. 2014. Impact of precipitation and runoff on ephemeral gully development in cultivated croplands. In: IAHS-AISH Proceedings and Reports. New Orleans: Copernicus GmbH, pp. 87-92. https://doi.org/10.5194/PIAHS-367-87-2015.

Liu, G., Zheng, F., Wilson, G. V., Xu, X. and Liu, C. 2021. Three decades of ephemeral gully erosion studies. *Soil and Tillage Research* 212. https://doi.org/10.1016/J.STILL.2021.105046.

Luquin, E., Campo-Bescós, M. A., Muñoz-Carpena, R., Bingner, R. L., Cruse, R. M., Momm, H. G., Wells, R. R. and Casalí, J. 2021. Model prediction capacity of ephemeral gully evolution in conservation tillage systems. *Earth Surface Processes and Landforms* 46(10), 1909–1925. https://doi.org/10.1002/ESP.5134.

Momm, H. G., Wells, R. R. and Bennett, S. J. 2018. Disaggregating soil erosion processes within an evolving experimental landscape. *Earth Surface Processes and Landforms* 43, 543–552. https://doi.org/10.1002/esp.4268.

Morgan, S., Huffaker, R., Giménez, R., Campo-Bescós, M. A., Muñoz-Carpena, R. and Govers, G. 2022. Experimental evidence that rill-bed morphology is governed by emergent nonlinear spatial dynamics. *Scientific Reports* 12(1), 21500.

Nachtergaele, J. and Poesen, J. 1999. Assessment of soil losses by ephemeral gully erosion using high-altitude (stereo) aerial photographs. *Earth Surface Processes and Landforms* 24, 693–706. https://doi.org/10.1002/(SICI)1096-9837(199908)24:8.

Nachtergaele, J., Poesen, J., Sidorchuk, A. and Torri, D. 2002. Prediction of concentrated flow width in ephemeral gully channels. *Hydrological Processes* 16(10), 1935–1953. https://doi.org/10.1002/HYP.392.

Nearing, M. A., Norton, L. D., Bulgakov,D. A. and Larionov, G. A. 1997. Hydraulics and erosion in eroding rills. *Water Resources Research* 33(4), 865–876.

Palmeri, V., di Stefano, C. and Ferro, V. 2021. A Maximizing Hydraulic Radius (MHR) method for defining cross-section limits in rills and ephemeral gullies. *Catena* 203, 105347. https://doi.org/10.1016/J.CATENA.2021.105347.

Parker, C., Clifford, N. J. and Thorne, C. R. 2012. Automatic delineation of functional river reach boundaries for river research and applications. *River Research and Applications* 28(10), 1708–1725. https://doi.org/10.1002/RRA.1568.

Poesen, J. and Govers, G. 1990. Gully erosion in the loam belt of Belgium: Typology and control measures. In: Boardman, J., Foster, I. D. L. and Dearing, J. A. (Eds.), *Soil Erosion on Agricultural Land*. London: John Wiley & Sons, pp. 513–530.

Poesen, J. and Valentin, C. 2003. Gully erosion and global change. *Catena* 50(2–4), 87–89. https://doi.org/10.1016/S0341-8162(02)00146-7.

Poesen, J. W. A. and Hooke, J. M. 1997. Erosion, flooding and channel management in Mediterranean environments of southern Europe. *Progress in Physical Geography: Earth and Environment* 21(2), 157–199. https://doi.org/10.1177/030913339702100201.

Poesen, J., Nachtergaele, J., Verstraeten, G. and Valentin, C. 2003. Gully erosion and environmental change: Importance and research needs. *Catena* 50(2–4), 91–133. https://doi.org/10.1016/S0341-8162(02)00143-1.

Poesen, J., Vanwalleghem, T., de Vente, J., Knapen, A., Verstraeten, G. and Martínez-Casasnovas, J. A. 2006. Gully erosion in Europe. In: Boardman, J. and Poesen, J. (Eds.), *Soil Erosion in Europe*. John Wiley and Sons Ltd: England, pp. 479–488.

Renard, K. G., Foster, G. R., Weesies, G. A., McCool, D. K. and Yoder, D. C. 1997. *Predicting Soil Erosion by Water: A Guide to Conservation Planning With the Revised Universal Soil Loss Equation (RUSLE) | Enhanced Reader*. Washington: U.S. Government Printing Office.

Ries, J. B. and Marzolff, I. 2003. Monitoring of gully erosion in the Central Ebro Basin by large-scale aerial photography taken from a remotely controlled blimp. *Catena* 50(2–4), 309–328. https://doi.org/10.1016/S0341-8162(02)00133-9.

Rock, G., Ries, J. B. and Udelhoven, T. 2012. Sensitivity analysis of uav-photogrammetry for creating Digital Elevation Models (DEM). *The International Archives of the Photogrammetry, Remote Sensing and Spatial Information Sciences* XXXVIII-1/C22, 69. https://doi.org/10.5194/ISPRSARCHIVES-XXXVIII-1-C22-69-2011.

Sancho, C., Benito, G. and Gutiérrez, M. 1991. *Agujas de Erosión y Perfiladores Microtopográficos*. Logroño: Geoforma Ediciones.

Seeger, M., Marzolff, I. and Ries, J. B. 2009. Identification of gully-development processes in semi-arid NE-Spain. *Zeitschrift fur Geomorphologie* 53(4), 417–431. https://doi.org/10.1127/0372-8854/2009/0053-0417.

Smith, L. M. 1993. Investigation of ephemeral gullies in Loessial soils in Mississippi, US army corps of engineers. Technical Report GL-93-11.

Soil Science Society of America. 1984. *Glossary of Soil Science Terms*. Madison, ed.

Takken, I., Govers, G., Ciesiolka, C. A. A., Silburn, D. M. and Loch, R. J. 1998. Factors influencing the velocity-discharge relationship in rills. Proceedings of the International Symposium on Modelling Soil Erosion, Sediment Transport and Closely Related Hydrological Processes, Vienna 1998. IAHS Publication Number 249, 63–70.

Thwaites, R. N., Brooks, A. P., Pietsch, T. J. and Spencer, J. R. 2022. What type of gully is that? The need for a classification of gullies. *Earth Surface Processes and Landforms* 47(1), 109–128. https://doi.org/10.1002/ESP.5291.

USDA-NRCS. 1997. *America's Private Land: A Geography of Hope*. Diane Publishing Company: Philadelphia.

Vanmaercke, M., Panagos, P., Vanwalleghem, T., Hayas, A., Foerster, S., Borrelli, P., Rossi, M., Torri, D., Casali, J., Borselli, L., Vigiak, O., Maerker, M., Haregeweyn, N., de Geeter, S., Zgłobicki, W., Bielders, C., Cerdà, A., Conoscenti, C., de Figueiredo, T., Evans, B., Golosov, V., Ionita, I., Karydas, C., Kertész, A., Krása, J., le Bouteiller, C., Radoane, M., Ristić, R., Rousseva, S., Stankoviansky, M., Stolte, J., Stolz, C., Bartley, R., Wilkinson, S., Jarihani, B. and Poesen, J. 2021. Measuring, modelling and managing gully erosion at large scales: A state of the art. *Earth-Science Reviews* 218. https://doi.org/10.1016/J.EARSCIREV.2021.103637.

Wells, R. R., Alonso, C. V. and Bennett, S. J. 2009a. Morphodynamics of headcut development and soil erosion in upland concentrated flows. *Soil Science Society of America Journal* 73(2), 521–530. https://doi.org/10.2136/SSSAJ2008.0007.

Wells, R. R., Bennett, S. J. and Alonso, C. V. 2009b. Effect of soil texture, tailwater height, and pore-water pressure on the morphodynamics of migrating headcuts in upland concentrated flows. *Earth Surface Processes and Landforms* 34(14), 1867–1877. https://doi.org/10.1002/ESP.1871.

Wells, R. R., Momm, H. G., Bennett, S. J., Bingner, R. L. and Dabney, S. M. 2011. An experimental study of gully sidewall expansion. ASABE – International Symposium on Erosion and Landscape Evolution 2011, 333–341. https://doi.org/10.13031/2013.39246.

Wu, S., Chen, L., Wang, N., Yu, M. and Assouline, S. 2018. Modeling rainfall-runoff and soil erosion processes on hillslopes with complex rill network planform. *Water Resources Research* 54, 10117–10133. https://doi.org/10.1029/2018WR023837.

Chapter 5

Advances in tracking sediment transport from agricultural land

Anthony J. Parsons, University of Sheffield, UK

1 Introduction

Almost 30 years ago, Evans and Boardman (1994) in an assessment of soil erosion by water in farmers' field in the UK recommended that 'Emphasis should be given to measuring the distance particles are transported by splash, wash and rills in an erosion event'. The distance particles are transported is important for several reasons. First, this distance is necessary for reconciling measurements of erosion made at different scales (see Parsons et al., 2004). Secondly, it impacts on travel times of pollutants that adhere to eroded soil particles. Thirdly, it is important for identifying sources of pollutants in waterways. Without knowing the rate at which soil particles move through the environment, it is impossible to relate the existence of adhered pollutants to present or past land-use practices. Tracking the movement of soil particles is, however, inherently difficult because of their size, typically well under 2 mm. Resolving this problem usually involves one of two compromises. Either monitoring the movement of individual particles is sacrificed in favour of gaining insight into the mean and statistical distribution of bulk sediment transport, or the movement of particles larger than typical soil particles is monitored in the hope that their movement can be related to the transport of typical soil particles. Only in rare instances has the movement of individual particles with size comparable to that of typical soils been studied. In this survey, I will report on these three types of advances into tracking sediment transport. In each section, I will consider tracing sediment

http://dx.doi.org/10.19103/AS.2023.0131.08

transport by splash, interrill flow (wash), rills and gullies where relevant studies have been undertaken. The survey will be confined to addressing the issue raised by Evans and Boardman and will not cover the use of tracers to estimate erosion rates such as the use of fallout and cosmogenic radionuclides. Nor will it cover those studies where the experimental design involved artificial tracer particles being applied to the whole surface such that sources and sinks of particles cannot be identified other than as areas of erosion and deposition in a manner comparable to the use of radionuclides (e.g. Ventura et al., 2001; Guzman et al., 2013; Liu et al., 2018).

2 Tracking bulk sediment transport

In a very early study, De Ploey and Savat (1968) used radioactive sand to study splash erosion which, they demonstrated, caused expansion of convex portions of hillslopes at the expense of other contiguous portions. Less environmentally detrimental, Parsons et al. (1993) used magnetite crushed to particles 80 μm in diameter (a compromise between an equivalent mass to the median soil particle size at the site and the effect of particle size on detachment rate) on an experimental plot in Arizona to study sediment transport in interrill flow. In this study, three artificial rainstorms at intensities of approximately 80 mm/h were applied to the plot for around 1 h. The results showed that sediment in interrill flow typically travels only a short distance during a single storm event - 29.7% of the transported sediment was deposited within 25 cm of its source during the first of the three experiments, and only 2.2% reached as far as the second survey line 2.95 m downslope of the seedline. The rapid decline in the amount of transported sediment with distance is consistent with an exponential distribution of travel distance both recorded in other studies of sediment transport (e.g. Hubbell and Sayre, 1964; Wainwright and Thornes, 1991) and argued for on theoretical grounds by Kirkby (1991). In a subsequent follow-up study at the same site (Parsons et al., 2010), it was shown that natural rainfall in the 16 years following the experiments had resulted in median transport of a further 0.18 m. Using this result, together with total sediment lost from the plot during the earlier rainfall simulation experiments, these authors were able to use the magnetite tracers to arrive at an estimate of interrill erosion of 0.02 mm/year.

A second approach to exploit magnetic tracing is to use heat treatment to enhance the magnetic susceptibility of naturally occurring soil particles. Although Parsons et al. (1993) found that such treatment did not achieve a sufficient increase in magnetic susceptibility to make tracking of the treated particles possible, Armstrong et al. (2012) were able to employ this approach to study the movement of a range of soil size fractions. These size fractions were placed in 5-mm-wide by 30-mm-deep trenches created by excavating the soil 40 mm from the top of 250-mm-long Perspex boxes. The soil boxes were subject to simulated rainfall at 52 mm/h for 1 h, and the magnetic susceptibility of the surfaces was mapped before and after the rainfall experiment. The results showed that between 2% and 22% of the soil eroded from the boxes originated from the trenches with, as expected, increased travel distances for finer particles.

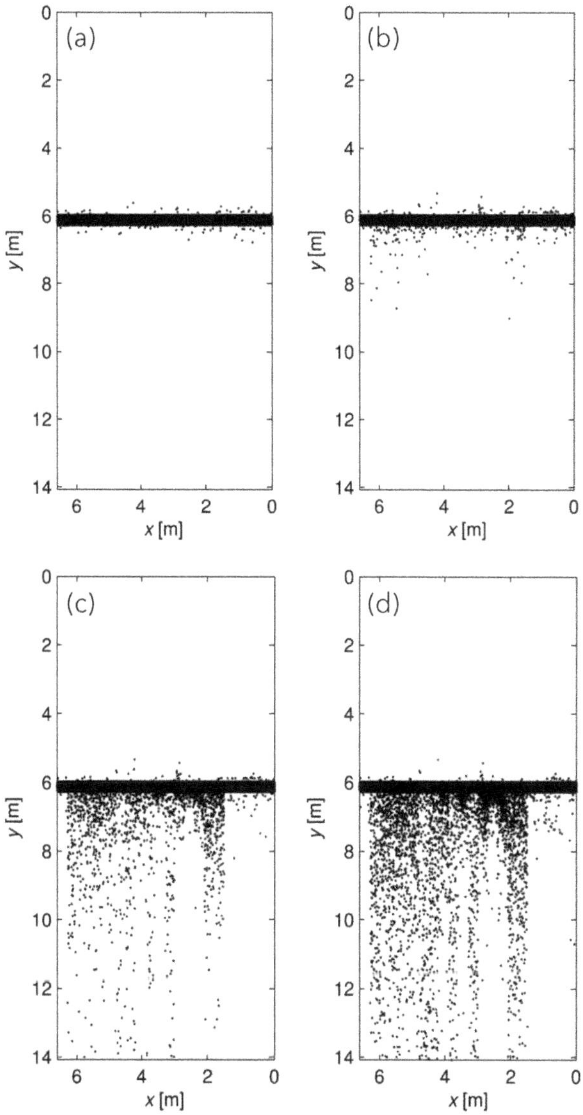

Figure 1 Modelling of the movement of Cs-137 particles at (a) 50 s, (b) 150 s, (c) 500 s and (d) 1200 s during a simulated rainfall event (after Cooper et al., 2012).

At a much larger scale, Cooper et al. (2012) used a similar experimental design in which Cs-137-rich soil was placed in a 0.3-m-wide trench located 5.7 m from the top of a plot 13.75-m long by 6.5-m wide which was subject to a 20-min simulated rainstorm at a design rainfall intensity of 60 mm/h. After the experiment, the plot was sampled to determine the loss of Cs-137 from the source line, its deposition on the plot and presence in the sediment leaving the plot. These authors used a marker-in-cell version of MAHLERAN (Wainwright et al., 2008) to predict the spatial pattern of the Cs-137 markers on the plot at various times during the experiment (Fig. 1), compared the predicted spatial pattern with that observed (Fig. 2) and the timing of the arrival of the Cs-137 at the plot outlet with the observed sediment delivery (Fig. 3).

A tracer that has been used in a number of studies is rare earths. Zhang et al. (2003, 2018) used five rare earths placed in 0.15-m bands in a 4-m-long soil box (Fig. 4) which was subjected to six rainfall-simulation experiments each lasting for 1 h and at intensities of 60 mm/h for the first four experiments and 90 mm/h for the final two. These authors were able to calculate the deposition of each rare earth along the slope after each rainfall simulation (Fig. 5) from which they inferred that sediment detachment, transport and deposition occurred simultaneously within the soil box and that the upper part was dominated by

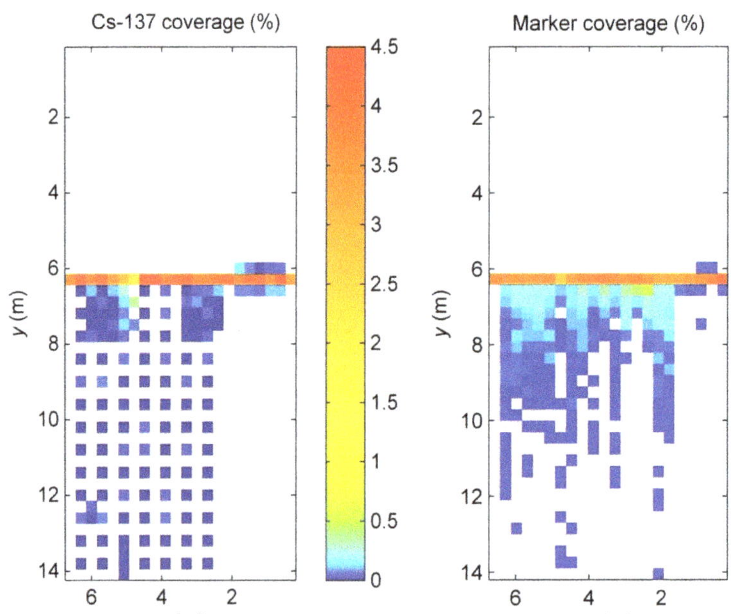

Figure 2 Comparison of (a) observed distribution of Cs-137 and (b) modelled markers (after Cooper et al., 2012).

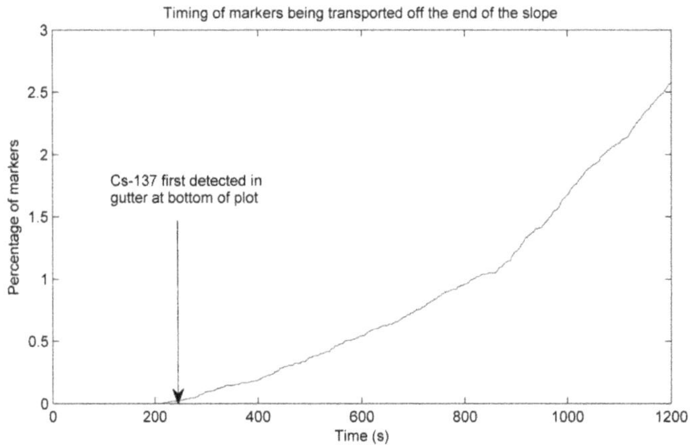

Figure 3 Comparison of modelled markers leaving the plot and timing of first recorded Cs-137 in eroded sediment (after Cooper et al., 2012).

erosion due to limited sediment influx whereas the lower section was limited by transport and largely acted as a conduit for sediment delivered from upslope. Their results were also interpreted to shed light on the mechanisms of sediment transport. They inferred that rainfall-driven rolling and creeping dominated the upper part of the plot, whereas flow-driven rolling and sliding became more important in the lower section.

Liu et al. (2004) used a similar experimental design in three sets of experiments – a small (6.5-m long) laboratory plot subject to rainfall simulation, a 1/75 scale model of a 0.18 km^2 gullied watershed, and a 100-m-long field plot for which runoff was measured from natural rainfall events over a 4-year period. In all cases, runoff and sediment were collected only from the outlets of the study plots, but the presence of the different rare earths gave information on the sources of sediment. These authors used this information to determine erosion rates along the length of the laboratory plot (Fig. 6), but also for the 100-m plot showed that between 25-4 and 35.8% of the two most upslope rare earths were deposited on the hillslope during low-intensity rainstorms, but that this figure fell to less than 10% of the eroded material in high-intensity events.

Zhang et al. (2018) also studied a model of a watershed using eight rare earths, each used to label a different section of the watershed (upstream upper slope, upstream lower slope, downstream upper slope, downstream lower slope, upper main rill, lower main rill, upper branch rill and lower branch rill). The model watershed was subjected to three rainfall simulations of 30-min duration at intensities of 60, 90 and 120 mm/h. Results showed most eroded soil came from the rills (on average 68% over the three experiments), a result that is consistent with Govers and Poesen (1988). However, unless the rills are re-supplied with sediment or grow constantly, such a proportion

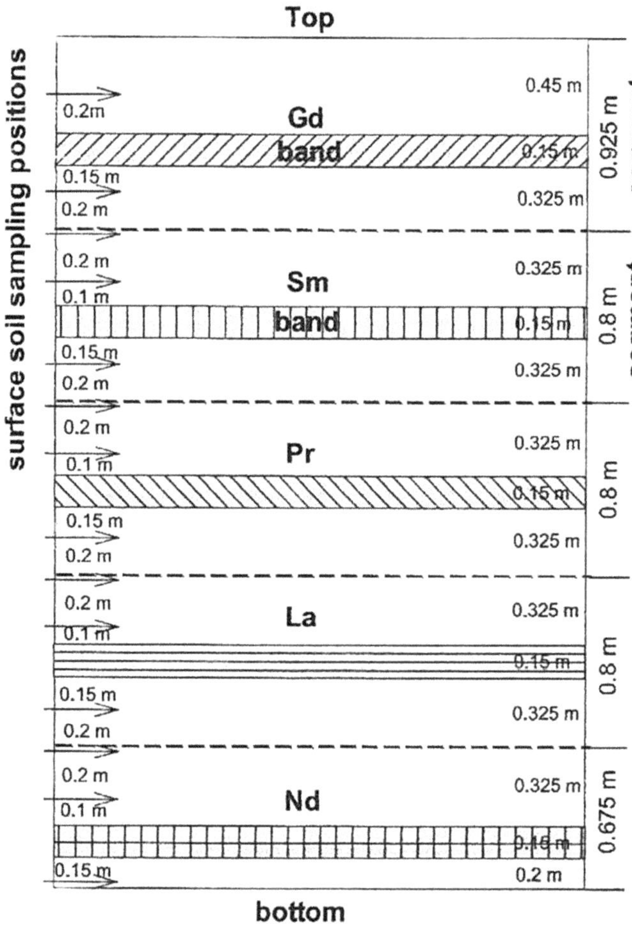

Figure 4 Schematic layout of tracked segments, locations of tracer band placements and soil surface sampling positions (after Zhang et al., 2017).

is unsustainable. Such is the case for the study by Michaelides et al. (2010) who used 10 rare earths in five rows spanning a break in slope from 15° to 2° in 6-m-long laboratory plot to investigate both rill and interrill sediment transport. Most sediment leaving the plot derived from the evolving rill network, but median transport distances for all of the rare earths range from 1 m to 2 m. As Liu et al. (2004) in the previously reported study found, the source of sediment varies with rainfall intensity. Likewise, Luk et al. (1993) found that whereas long-duration rainfall events (at high intensity) resulted in rill scour, short-duration events caused net filling of rills by sediment derived from interrill areas.

Stephens and Quinton (2008) used a similar experimental design to examine the movement of sediment into a longitudinal (downslope) tramline

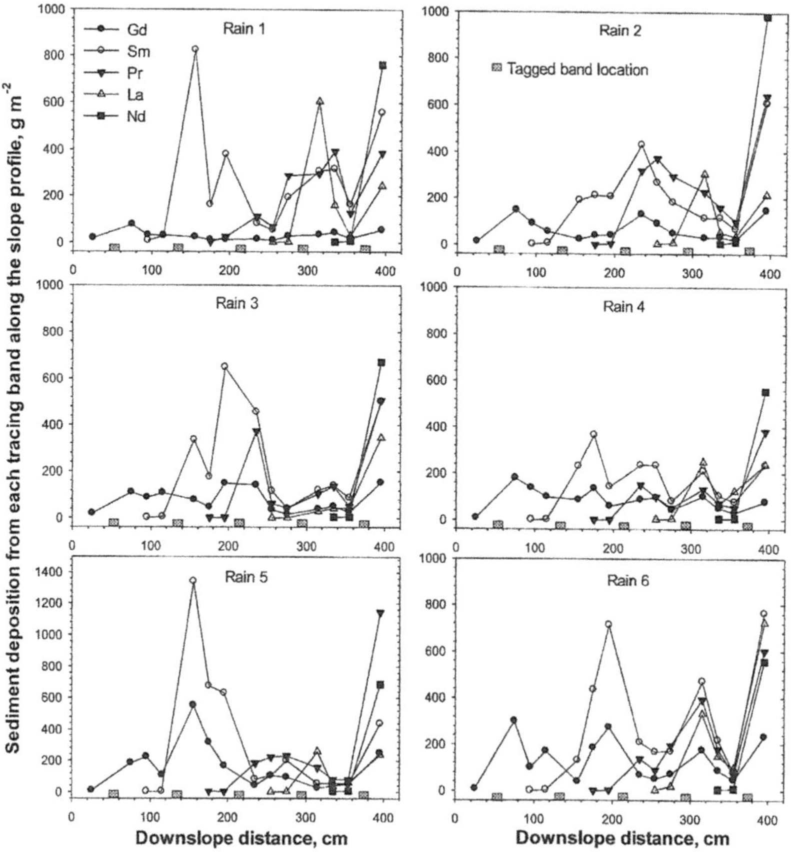

Figure 5 Sediment deposition rates from each tracer band along the slope following each of the six rainfall events (after Zhang et al., 2017).

in a field sown to oats. During the 2-week period of study, 25.6 mm of rainfall occurred. The study showed that most soil entering the tramline was derived from less than 1 m away, but some came from the most distant of the rare earth seedlines. This result is consistent with the findings of Parsons et al. (1993) who showed that in interrill areas most sediment travels only a short distance.

Polyakov et al. (2004) applied six rare earths to a small catchment planted to soybeans and monitored it over a 6-month period. Their observations showed that sediment either moved a short distance in interrill areas or was able to leave the catchment if it was transported into channels. Kimoto et al. (2006) studied the same watershed over a 4-year period during which vegetation cover included both soybeans, wheat and clover. These authors concluded that splash and interrill erosion provided the dominant source of sediment leaving the catchment in small runoff events and that the importance of rill and channel sources increased with storm size. However, the authors recognised

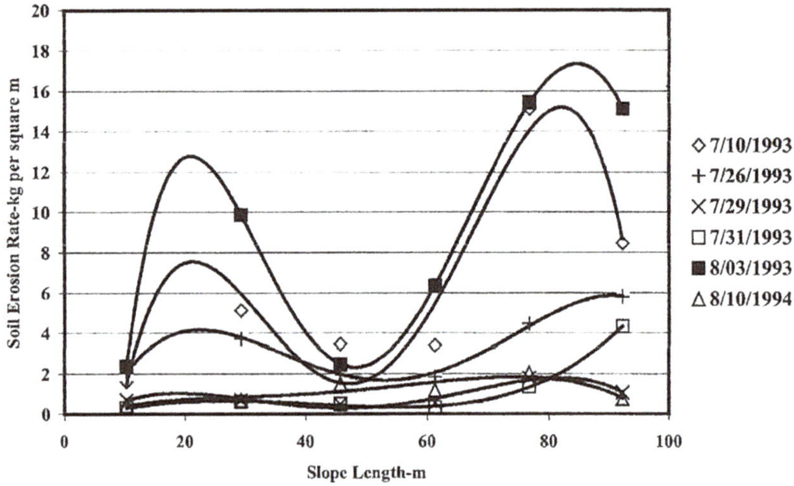

Figure 6 Soil erosion rate (kg/m^2) versus distance downslope for different dates in field soil erosion experiment (after Liu et al., 2004).

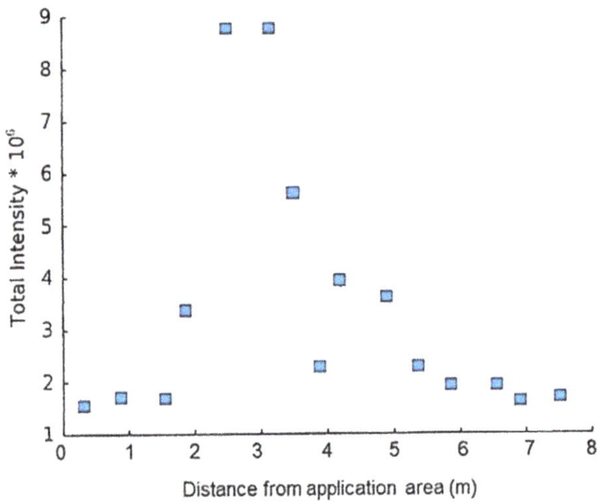

Figure 7 Distribution of tracer particles after tillage operations (after Hardy et al., 2019).

that, because they collected sediment only from the catchment outlet, it was impossible to determine the event source of sediment, only its ultimate origin. In that sense, this study is analogous to studies using sediment fingerprinting.

An advance in bulk sediment tracing is the study by Hardy et al. (2019), who used quartz particles coated with a fluorescent pigment in a field study to examine the movement of particles in response to tillage. The particles were placed in a trench 1 m by 0.3 m and 0.3-m deep located 12 m from the top of the study plot which was then subject to seven tillage operations. Images were then taken of the plot starting 5 m upslope of the tracer location and extending 10 m downslope of it. Using fluorescent intensity as a surrogate for numbers of particles, results showed the technique is able to record the distribution of particle travel distance (Fig. 7). This tracer also has the very important capability to record sediment transport in real time (Hardy et al., 2016), thus enabling the monitoring of sediment movement during a rainfall event.

3 Tracking individual particles

To track the movement of individual particles, Parsons et al. (2014, 2018) developed simulated soil particles embedded with radio-frequency identification (RFID) tags. The tags were encased in clear silicone resin and then coated with bronze powder to give them a density comparable to that of quartz particles. These surrogate soil particles were between 2.5 mm and 4.0 mm in diameter and had specific weights of 2.5 to 3.0. Three experiments were performed using these particles; a laboratory study in which all movement of the particles occurred due to raindrop impact, a field study in which particle movement was measured on five occasions over seven rainfall events during a 4-month period on a bare soil plot, and a laboratory study in which the surface was covered in quartz grains of similar (2.86 mm) median diameter and subject to a rainfall simulation event long enough for equilibrium runoff to be attained. The raindrop-impact study showed a distribution of particle movement best approximated by a Dagum 4-parameter distribution closely allied to the distribution found by Long et al. (2014) - see later. The field study showed that, despite the diverse topographic locations of the particles, there is an increase in particle movement with distance downslope that is well correlated with total runoff and even more closely with peak runoff prior to the survey (Fig. 8). The second laboratory experiment revealed a mean distance of particle movement of 125.7 mm and a median travel distance of 73.1 mm (Fig. 9).

Probably, the most sophisticated form of individual particle tracking is particle imaging velocimetry (PIV). Long et al. (2014) used PIV to study particle detachment and transport by raindrop impact (Fig. 10). The study showed a hitherto unreported identification of two distinct mechanisms of raindrop detachment and transport: the ejection of water-encapsulated particles from the edge of the region wetted by the raindrop and the ejection of dry particles from the periphery of the impact crater. The water encapsulated particles are few in number, have low ejection angles and have high ejection velocities leading to large transport distances. In contrast, the ejected dry particles are much greater in number, are characterised by high ejection angles and have low ejection velocities. Hence they have lower transport distances. Significantly, the study showed that only 2% of the water-droplet momentum at impact is

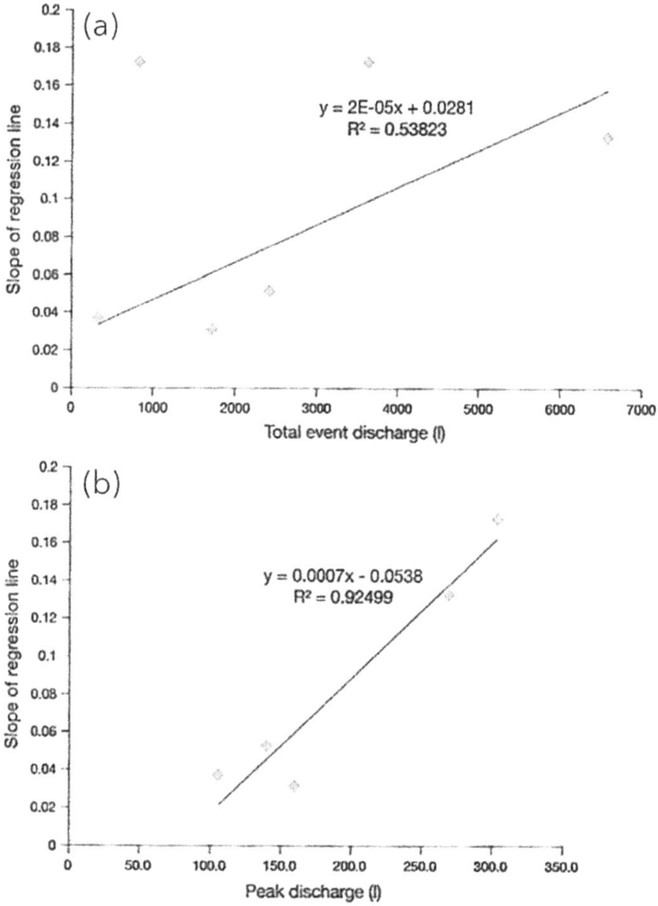

Figure 8 Relationship between the distance of movement of RFID-tagged particles and (a) total runoff prior to the survey and (b) peak runoff prior to the survey (after Parsons et al., 2014).

transferred to the ejected particles, a much lower value than might be expected from previous research (Mihara, 1951). This study was, however, conducted on a dry bed of sand with a size range of 150–160 μm. How these results might differ for natural soil with a much wider range of particle sizes and where the surface was wetted by multiple raindrops requires further investigation.

Hardy et al. (2017) used a similar approach to track the movement of individual fluorescent particles during simulated rainfall (Fig. 11). The particles were 250 μm in diameter and placed on the surface of a sandy loam soil in a soil box 0.2 m × 0.3 m × 0.15 m subjected to rainfall lasting 4 min at an intensity of 40 mm/h.

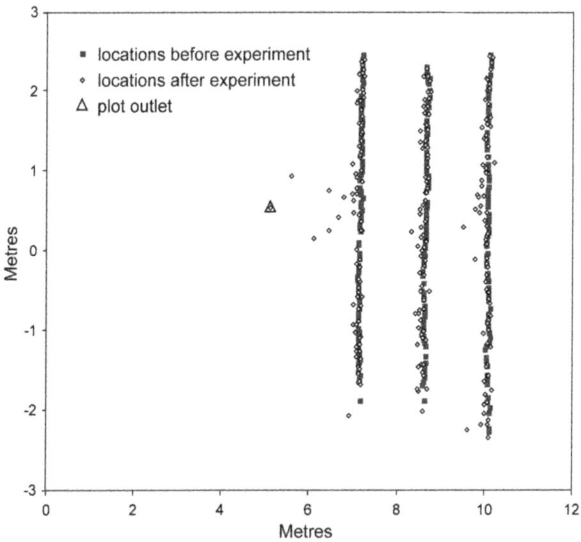

Figure 9 Distribution of RFID-marked particles before and after rainfall simulation (after Parsons et al., 2018).

4 Understanding sediment transport

The majority of the studies reported in this chapter have been conducted on small, bare laboratory plots. Few studies have been conducted at the scale of agricultural fields, or with a vegetation cover. Consequently, their interpretation in terms of erosion of agricultural soils is subject to some provisos. All tracer particles need to be introduced to the soil surface. Guzman et al. (2013, p 132) note that binding the powdered magnetite to the soil requires the soil to be disturbed. Parsons et al. (1993), who simply spread magnetite on the (partly vegetated) soil surface found that more magnetite was moved in the first experiment than during subsequent ones. In all likelihood, this difference was a result of the greater incorporation of the magnetite into the surface soil layer with time. It is probably fair to suggest that the rates of tracer movement represent something close to the maximum that might be observed from agricultural areas. Typically, laboratory studies have used high-intensity rainfall simulation. As some of the studies have shown, sources of sediment appear to vary with rainfall intensity (Liu et al., 2004; Michaelides et al., 2010) or event magnitude (Luk et al., 1993)

Notwithstanding these comments, the studies do show that particles move relatively short distances during single (or even multiple) storm events. The implications of this finding are that (1) scaling up of laboratory experiments to agricultural areas will not straightforward and (2) sediment moves slowly through the landscape. This latter implication has considerable significance

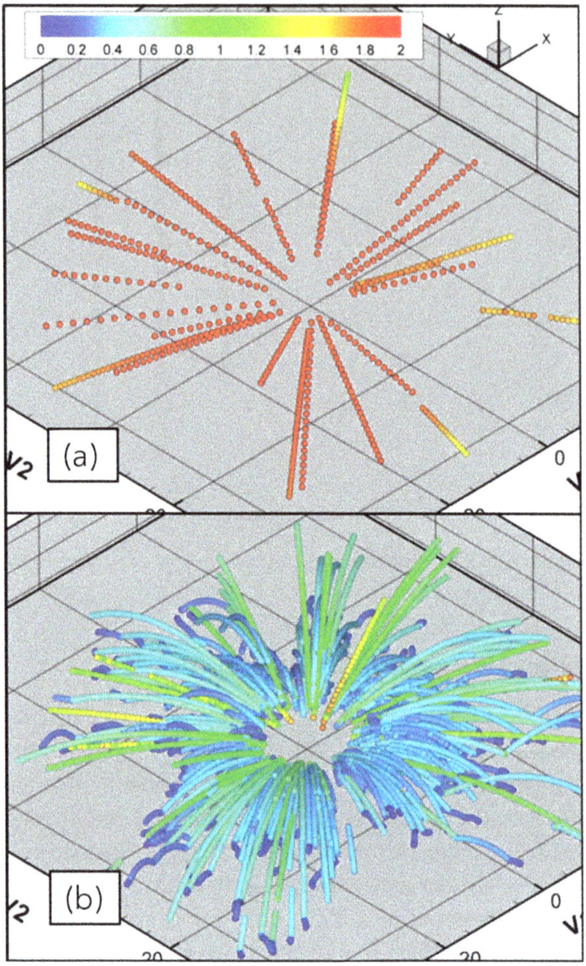

Figure 10 Particle location tracks according to particle type: (a) high velocity, low ejection angle, high displacement and (b) low velocity, high ejection angle, low displacement (after Long et al., 2014).

for pollution monitoring. Tracer studies have also shed light on the mechanics of sediment transport. In interrill areas, where sediment transport is achieved by a combination of raindrop detachment and flow transport, sediment moves by a combination of rolling, creeping and sliding. Furthermore, some evidence indicates sources of sediment appear to be upslope areas, with lower slope areas contributing mainly as conduits of this sediment. Such an idea is already crudely incorporated in many soil erosion models via the concept of transport capacity, whereby the rate of sediment detachment is a function of the proportion of transport capacity already satisfied (e.g. Nearing et al., 1989;

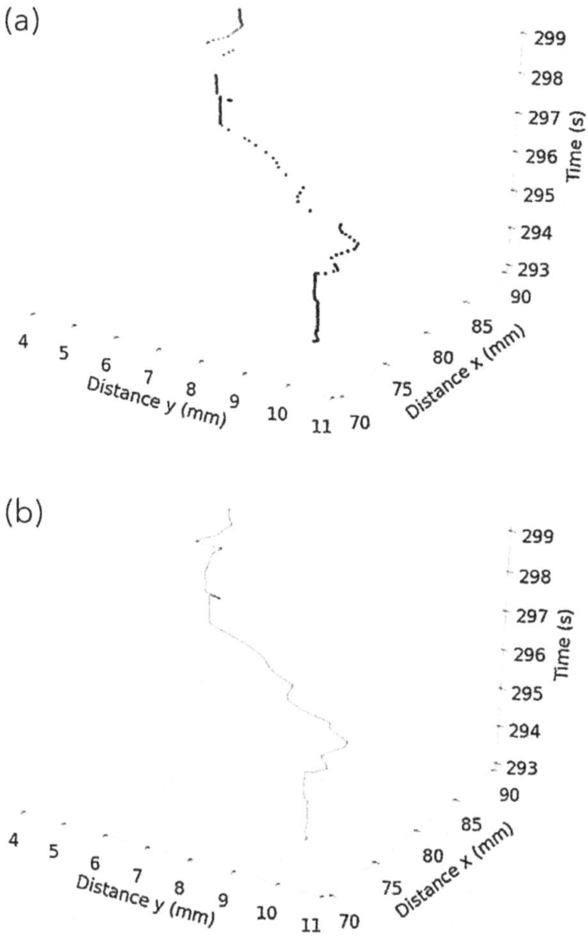

Figure 11 The movement of a single particle over 218 frames (4.36 s). (a) The precise locations each time the particle was detected. (b) The pathway created by interpolating between the locations given in (a). The particle was detected in all but one of the 218 frames (after Hardy et al., 2017).

Morgan et al., 1998). But such a crude approximation makes the assumption that sediment already detached remains in transport. This assumption is at odds with the tracer experiments showing short travel distances for individual particles.

5 Conclusion

Studies tracking the movement of individual particles have done much to meet the recommendation of Evans and Boardman (1994). However, most studies are small scale, and further work is needed to implement their findings to erosion from agricultural land. In combination with models that explicitly consider transport distance (e.g. Wainwright et al., 2008), they should enable better management of measures to mitigate erosion and pollution transfers.

6 Where to look for further information

- https://geology.com/articles/rare-earth-elements/.
- https://gmw.com/magnetic-susceptibility/.
- https://www.atlasrfidstore.com/rfid-beginners-guide/.
- Raffel, M., Willert, C. E., Werely, S. T. and Kompenhans, J. 2007. *Particle Image Velocimetry*. Springer, Berlin.

7 References

Armstrong, A., Quinton, J. N. and Maher, B. A. 2012. Thermal enhancement of natural magnetism as a tool for tracing eroded soil. *Earth Surface Processes and Landforms* 37(14), 1567–72.

Cooper, J. R., Wainwright, J., Parsons, A. J., Onda, Y., Fukuwara, T., Obana, E., Kitchener, B., Long, E. J. and Hargrave, G. H. 2012. A new approach for simulating the distribution of soil particles by water erosion: a marker-in-cell model. *Water Resources. Research* 117, F04027, DOI: 10.1029/2012JF002499.

De Ploey, J. and Savat, J. 1968. Contribution a l'étude de l'érosion par le splash. *Zeitschrift für Geomorphologie N.F* 12, 174–93.

Evans, R. and Boardman, J. 1994. Assessment of water erosion in farmers' fields in the UK. In: *Conserving Soil Resources: European Perspectives* Rickson, J. (Ed.). CAB International: Wallingford, pp. 13–24.

Govers, G. and Poesen, J. 1988. Assessment of the interrill and rill contributions to total soil loss from an upland field plot. *Geomorphology* 1(4), 343–54.

Guzman, G., Quinton, J. N., Nearing, M. A., Mabit, L. and Gómez, J. A. 2013. Sediment tracers in water erosion studies: current approaches and challenges. *Journal of Soils and Sediments* 13(4), 816–33. DOI: 10.1007/s11368-013-0659-5.

Hardy, R. A., Pates, J. M., Quinton, J. N. and Coogan, M. P. 2016. A novel fluorescent tracer for real-time tracing of clay transport over soil surfaces. *CATENA* 141, 39–45.

Hardy, R. A., James, M. R., Pates, J. M. and Quinton, J. N. 2017. Using real time particle tracking to understand soil particle movements during rainfall events. *CATENA* 150, 32–8. DOI: 10.1016/j.catena.2016.11.005.

Hardy, R. A., Quinton, J. N., James, M. R., Fiener, P. and Pates, J. M. 2019. High precision tracing of soil and sediment movement using fluorescent tracers at hillslope scale. *Earth Surface Processes and Landforms* 44(5), 1091–99.

Hubbell, D. W. and Sayre, W. W. 1964. Sand transport studies with radioactive tracers. *Journal of the Hydraulics Division* 90(3), 39–68.

Kimoto, A., Nearing, M. A., Shipitalo, M. J. and Polyakov, V. O. 2006. Multiyear tracking of sediment sources in a small agricultural watershed using rare earth elements. *Earth Surface Processes and Landforms* 31(14), 1763–74.

Kirkby, M. J. 1991. Sediment travel distance as an experimental and model variable in particulate movement. *Catena. Supplement* 19, 111–28.

Liu, P.-L., Tian, J.-L., Zhou, P.-H., Yang, M.-Y. and Shi, H. 2004. Stable rare earth element tracers to evaluate soil erosion. *Soil and Tillage Research* 76(2), 147–55.

Liu, L., Huang, M., Zhang, K., Zhang, Z. and Yu, Y. 2018. Preliminary experiments to assess the effectiveness of magnetite powder as an erosion tracer on the Loess Plateau. *Geoderma* 310, 249–56.

Long, E. J., Hargrave, G. K., Cooper, J. R., Kitchener, B. G. B., Parsons, A. J., Hewett, C. and Wainwright, J. 2014. Experimental investigation into the impact of a liquid droplet onto a granular bed using 3D, time-resolved, particle tracking. *Physical Revue. Part E* 89(3), 032201-1-14. DOI: 10.1103/PhysRevE.00.002200.

Luk, S.-H., Abrahams, A. D. and Parsons, A. J. 1993. Sediment sources and sediment transport by rill flow and interrill flow on a semi-arid piedmont hillslope, southern Arizona. *catena* 20, 93–111.

Michaelides, K., Ibraim, L., Nord, G. and Esteves, M. 2010. Tracing sediment redistribution across a break in slope using rare earth elements. *Earth Surface Processes and Landforms* 35, 575–87.

Mihara, Y. 1951. Raindrops and soil erosion. Bulletin No 1. *National Institution of Agricultural Science*, Toyko, Japan, 59 pp.

Morgan, R. P. C., Quinton, J. N., Smith, R. E., Govers, G., Poesen, J. W. A., Auerswald, K., Chisci, G., Torri, D. and Styczen, M. E. 1998. The European Soil Erosion Model (EUROSEM): a dynamic approach for predicting sediment transport from fields and small catchments. *Earth Surface Processes and Landforms* 23(6), 527–44.

Nearing, M. A., Foster, G. R., Lane, L. J. and Finkner, S. C. 1989. A process based soil erosion model for USDA water erosion prediction project technology. *Transactions of the American Society of for Agricultural Engineers* 32, 1587–93.

Parsons, A. J., Cooper, J., Onda, Y. and Sakai, N. 2018. Application of RFID to soil-erosion research. *Applied Sciences* 8(12), 2511, DOI: 10.3390/app8122511.

Parsons, A. J., Wainwright, J. and Abrahams, A. D. 1993. Tracing sediment movement in interrill overland flow on a semi-arid grassland hillslope using magnetic susceptibility. *Earth Surface Processes and Landforms* 18(8), 721–32.

Parsons, A. J., Wainwright, J., Fukuwara, T. and Onda, Y. 2010. Using sediment travel distance to estimate medium-term erosion rates: a 16-year record. *Earth Surface Processes and Landforms* 35(14), 1694–700. DOI: 10.1002/esp.2011.

Parsons, A. J., Onda, Y., Noguchi, T., Patin, J., Cooper, J., Wainwright, J. and Sakai, N. 2014. The use of RFID in soil-erosion research. *Earth Surface Processes and Landforms* 39, n/a-, DOI: 10.1002/esp.3628.

Polyakov, V. O., Nearing, M. A. and Shipitalo, M. J. 2004. Tracking sediment redistribution in a small watershed: implications for agro-landscape evolution. *Earth Surface Processes and Landforms* 29(10), 1275–91.

Stephens, C. J. and Quinton, J. N. 2008. Investigating source areas of eroded sediment transported in concentrated overland flow using rare earth elements. *CATENA* 74, 31–6.

Ventura, E., Nearing, M. A. and Norton, L. D. 2001. Developing a magnetic tracer to study soil erosion. *CATENA* 43(4), 277–91.

Wainwright, J. and Thornes, J. B. 1991. Computer and hardware modelling of archaeological sediment transport on hillslopes. In: *Computer Applications and*

Quantitative Techniques in Archaeology 1990, BAR International Series 565 Rahtz, S. and Lockyear, K. (Eds). BAR International: Oxford, pp. 183–94.

Wainwright, J., Parsons, A. J., Müller, E. N., Brazier, R. E., Powell, D. M. and Fenti, B. 2008. A transport-distance approach to scaling erosion rates: 1. Background and model development. *Earth Surface Processes and Landforms* 33(5), 813–26, DOI: 10.1002/esp.1624. Zhang et al. (2003, 2017).

Zhang, X. J., Liu, G. and Zheng, F. 2018. Understanding erosion processes using rare earth element tracers in a preformed interrill-rill system. *Science of the Total Environment* 625, 920–27.

Chapter 6

Advances in modeling soil erosion risk

Sudhanshu S Panda, University of North Georgia, USA; Debasmita Misra, University of Alaska Fairbanks, USA; Devendra M Amatya and Johnny M Grace III, USDA Forest Service, USA; and Anita Thompson, University of Wisconsin-Madison, USA

1 Introduction

Evidence of climate change and its increasing impact on the environment and the erosion of soil has caught the attention of researchers globally. Soil erosion, considered as one of the worst geohazards human facing on the earth, is one of the environmental phenomena most affected by climate change. In the scientific community, there is a consensus that a critical impact of climate change is the increase in the quantity and intensity of precipitation (Giang et al.,

http://dx.doi.org/10.19103/AS.2023.0131.09

2017). An increase in precipitation would lead to higher rates of soil erosion (Parry et al., 2007). Many studies (Zhang and Nearing, 2005; Zhang et al., 2005; Pruski and Nearing, 2002; Nearing, 2001; Giang et al., 2017; Almargo et al., 2017; Licciardello et al., 2019) examined the change of soil erosion in response to changing climate conditions using global climate models (GCMs) and confirmed an increase in the rate of soil erosion.

The United States Geological Survey (USGS) regional regression equations (Feaster et al., 2014; Wagner et al., 2016), the Rational Method (Kuichling, 1889), and the Soil Conservation Service-Curve Number (SCS-CN) method (NRCS, 2004) have been widely used to estimate design-storm peak discharge for different return intervals (ODOT, 2014). These methods are based on site-specific hydraulic characteristics that can be represented by simple watershed average parameters, i.e. runoff coefficient, curve number (CN), time of concentration, and basin characteristics (size, shape, topography, soils, geology, and land use) (Schwab et al., 1971).

Precipitation intensity-duration-frequency (PIDF) curves have been developed to take account of climatic factors (US Water Resource Council, 1981; Hrachowitz et al., 2013; Rogger et al., 2012). These PIDFs have connected the extreme precipitation storm event values that are now being observed worldwide (in the last few decades) with climate change (Amatya et al., 2021). The US National Oceanic and Atmospheric Administration (NOAA) Atlas 14-based gridded PIDF values (Bonnin et al., 2006; Perica et al., 2013) are available in raster format that uses regional frequency analysis using the L-moments method (Hosking and Wallis, 2005) with annual maximum and partial series storm events data from various weather stations managed by different government agencies in the United States. Although the precipitation data is not fully up-to-date, comprehensive historic records have been used in developing PIDFs (Panda et al., 2022). Researchers, engineers, and land managers in United States have the opportunity to make increasing use of climate change scenario storm (PIDF) data in soil erosion analysis (Panda et al., 2022).

Biassutti and Seager (2015) suggest that soil erosion has generated significant environmental, social and economic damage globally. The impact of climate change on soil erosion will severely affect all types of land cover, especially those with a steep topography. We are experiencing soil erosion in the form of landslides in many parts of the world as a consequence of climate change, with alternating La-Nina and El-Nino conditions (Moreiras et al., 2018). According to the United States Department of Agriculture (USDA) Forest Service (FS) (Rasmussen et al., 2018), nearly 200 million acres of public land are susceptible to a wide range of climate change impacts in every region of the country. The extreme rainfall events would cause potentially severe damage to the USDA-FS-managed road system in mountainous and less accessible terrains due to soil erosion (Panda et al., 2018, 2019; Amatya et al., 2017).

NOAA Atlas-14 sources probably do not use local data from long-term precipitation gauges on small headwater catchments located in several USDA-FS experimental forests and ranges (EFRs) (Amatya et al., 2019; Laseter et al., 2012; Moran et al., 2008). However, PIDF raster formatted data developed using surface interpolation algorithms for individual states of USA is quite useful in local soil erosion analysis (Panda et al., 2022). Developing modified soil erosion models using GCM supported by climate data are essential for watershed management and would pave way for identifying the best soil conservation measures for safeguarding many vulnerable engineering structures from soil erosion due to extreme weather events.

2 The development of the universal soil loss equation and revised universal soil loss equation models

USLE, the refined mathematical universal soil loss equation model, developed by Wischmeier and Smith (1965a,b), is the most commonly used model for predicting soil losses due to area erosion. The model is also useful in determining the adequacy of conservation measures in farm management planning and for predicting nonpoint sediment losses in pollution control programs. It is presented in algorithmic form as:

$$A = R \times K \times L \times S \times C \times P \tag{1}$$

where

$A =$	Computed soil loss per unit area
$R =$	Rainfall factor: the number of erosion index units in the period of consideration. This is the measure of erosive force of specific rain.
$K =$	Soil erodibility factor: the erosion rate/unit of erosion index for a specific soil, in a cultivated continuous fallow on a 9% slope and 22.1 m long.
$L =$	Slope length factor: the ratio of soil loss from field slope length to that from a 22.1 m length on the same soil type and gradient.
$S =$	Slope gradient factor: the ratio of soil loss from the field gradient to that from a 9% slope on the same soil type and slope length.
$C =$	Cropping management factor: the ratio of soil loss from a field with specific cropping and management to that from the fallow condition on which K is evaluated.
$P =$	Erosion control practice factor: the ratio of soil loss with contouring, strip cropping, or terracing to that with straight row farming, up and down slope.

In the 1940s, A. W. Zingg (Zingg, 1940) introduced the first mathematical equation to estimate soil erosion losses in a field experiment based on natural and simulated rainfall using a Shelby loam soil in Missouri (USA). The equation is:

$$X = CS^{m}L^{n} \quad (2)$$

where X is total soil loss from a land slope of unit width, C is a constant of variation, S is land slope (%), L is horizontal length of land slope (ft), and m and n are empirical exponents.

Zingg (1940) then defined average soil loss per unit area from a slope of unit width represented by:

$$A = CS^{m}L^{n-1} \quad (3)$$

where the values of m and n (derived from the simulated rainfall experiment) are 1.4 and 0.6, respectively.

Smith (1941) modified Zingg's soil loss equation with the introduction of a factor, P, representing the ratio of soil loss with a mechanical conservation practice with respect to soil loss without the practice. The new equation by Smith (1941) used the same values of exponent m and n as developed by Zingg (1940) and presented it as:

$$A = CS^{1.4}L^{0.6}P \quad (4)$$

Smith's work established the concept of an allowable soil loss, now known as the 'T-Value' (tolerable soil loss value), based on maintenance of soil fertility (Laflen and Flanagan, 2013). Browning et al. (1947) and Musgrave (1947) later introduced a broad range of factors to account for in erosion estimation such as crop production, soil characteristics, and climate condition, paving the way for a more complete soil erosion equation.

Wischmeier and Smith (1965a,b) subsequently made available the first published USLE predicting rainfall-erosion losses from cropland east of the Rocky Mountains. USLE has been updated three times as:

- USLE (Wischmeier and Smith, 1978);
- Revised USLE (RUSLE) (Renard et al., 1997); and
- RUSLE2 (USDA-ARS, 2013).

The USLE, RUSLE, and RUSLE2 models, all follow the same equational concept of multiplying soil erosion-inducing spatial factors such as R (rainfall erosivity factor), K (soil erodibility factor), L (slope length factor), S (slope gradient factor), C (cropping management factor), and P (erosion control management factor) to obtain a computed soil loss per unit area.

Among those six factors, the R-factor is particularly important given the increasing occurrence of extreme storm events. Along with the modification of the R-factor, L- and S-factors can be modified with greater accuracy because of the availability of LiDAR-based topographic data. The LiDAR-based digital elevation model (DEM) would improve the accuracy of the L- and S-factors to quantify soil erosion correctly. The advent of unmanned aerial vehicle

(UAV)-based image acquisition and improved image segmentation algorithms is also paving way for more accurate *C*- and *P*-factor development. This chapter reviews developing a modified RUSLE model using ArcGIS ModelBuilder automation to develop processes for improving each USLE factor so that soil erosion quantification can be estimated more precisely in both temporal and spatial dimensions.

USLE has been integrated with geographic information system (GIS) data to estimate the amount of soil erosion from a watershed since the mid-1990s. GIS data helps users manipulate and analyze spatial data with ease and accurately identify spatial locations vulnerable to soil erosion (Yitayew et al., 1999; Ouyang and Bartholic, 2001; Lufafa et al., 2003; Panda et al., 2004).

The revised universal soil loss equation (RUSLE), a newer version of USLE, was developed by modifying it to estimate *R*-, *K*-, *C*-, and *P*-factors, as well as soil erosion, more accurately (Renard et al., 1991). The use of the ARC Macro Language (AML) program was not sufficiently fast in 1990s to spatially calculate the *LS*-factor (Panda et al., 2004). Van Remortel et al. (2004) developed an array-based C++ program to automate the calculation of the *LS*-factor from the DEM.

However, with the availability of the latest high-spatial soil data (gSSURGO, 10 m), *L*- and *S*-factor development has become easier. The *L*- and *S*-factors are available in three soil attributes (low depth, high depth, and representative) format in the gSSURGO characteristics (attributes) database. With the advent of graphical user interface (GUI)-based GIS software and advances ModelBuilder technology in GIS, it has become easier to develop RUSLE models to determine soil erosion rate from a watershed. Many researchers have used RUSLE models to estimate the amount of erosion from fields and watersheds (Renard et al., 1997; Millward and Mersey, 1999). However, few have used ArcGIS ModelBuilder to build a spatial model using the RUSLE equation to estimate soil erosion from a watershed (Sharma, 2008; Daggupati et al., 2009; Panagopoulos and Ferreira, 2010; Panda et al., 2022). The RUSLE model developed in the ArcGIS ModelBuilder environment has helped to automate the process.

3 Introducing the *R*-factor (rainfall factor) into universal soil loss equation/revised universal soil loss equation/revised universal soil loss equation 2

Wischmeier (1959) used precipitation and soil loss data from fallow plots to assess the role of rainfall in estimating storm soil loss. The rainfall-runoff erosivity factor (*R*) was developed as a measure of the erosion force of a specific rainfall event. The *R*-factor is calculated as the average annual summation (EI_{30}) values of the amount of rainfall from storms with no less than 1.27 cm (0.5 inch) in a normal precipitation year or with a maximum 15-min intensity >25.4 mm h^{-1} (Wischmeier and Smith, 1978). In RUSLE, a longer storm period is divided into two storms when there is <0.13 cm (0.05 inch) during 6-h storm period (Yin et al., 2017). A monthly erosivity density map was developed in RUSLE2 using 15-min data from 3700 stations based on monthly precipitation data with a spatial resolution of 4 km obtained from the Parameter elevation Regression on Independent Slopes Model (PRISM, http://www.prism.oregonstate.edu/) database.

The *R*-factor represents the erosivity of climatic conditions at a particular location. Its average value has been determined from historical weather records. Wischmeier (1959) has evaluated the suitability of the *R*-factor at other locations (compared to the three experimental plot locations used in its development) for various cropping periods. The R-value proved to be effective for estimating rainfall-related soil loss (Laflen and Flanagan, 2013). By 1965 Isoerodent maps were available with the distribution of EI_{30} for half-month periods for areas east of the Rocky Mountains (Wischmeier and Smith, 1965b). The publication also provided statistics related to probabilities of occurrence of single storms and single-year *R*-values for many locations in the USA.

It has been assumed that storm losses from rainfall are directly proportional to the product of the total kinetic energy of the storm (E) multiplied by its maximum 30-min intensity (IWR, 2002). The erosivity of an individual storm is computed as the product of the storm's total energy (*E*), which is closely related to storm amount, and the storm's maximum 30-min intensity (I_{30}). The *R*-factor represents the average storm EI_{30} values over a year based on two key storm characteristics affecting erosivity: the amount of rainfall and peak intensity sustained over an extended period (IWR, 2002). However, with the exception of mean annual precipitation (P), I_{30} data is not readily available for most geographic areas. As a result, *R*-factor estimates as a function of *P* are often used since *P* is a parameter that is available for most geographic regions of the world.

Wischmeier and Smith (1965a) developed the following empirical equation for determining the energy (E) required to initiate the motion of sediment particles:

$$E = 916 + 331 \log_{10} I_{30} \tag{5}$$

where *E* is in foot-tons per inch (23.11 metric tons mm^{-1}) and *I* is the intensity of the storm in inches per hour (25.4 mm h^{-1}).

They found that, for a given storm, energy did not increase with intensity above 3 in h^{-1}, so the maximum value used for in this equation is 3 in h^{-1}. The *E* value was determined from the storm intensity (*I*) values and I_{30} values for each individual storm over the period of record. The product of *I* and I_{30} was calculated, summed, and averaged annually over the period of record to provide the *R*-factor as represented by:

$$R = \frac{1}{n}\sum_{j=1}^{n}\left[\sum_{k=1}^{m}(E)I_{30k}\right] \quad (6)$$

where k is the number of the individual storms to a total of m storms in a year, and j represents the year up to the total number of years, n, over which data was collected.

Renard et al. (1997) published the RUSLE with a new method to determine R-factor. However, they still used the EI parameter and the summation/averaging equation detailed above with I_{30} determination in the same manner as suggested by Wischmeier and Smith (1965a). However, E values were determined using an alternative equation given as:

$$E = \sum_{k=1}^{n} e_r v_r \quad (7)$$

where e_r is the rainfall energy per unit depth of rainfall per unit area (foot-tons per acre per inch = 9.25 metric tons ha^{-1} mm^{-1}), v_r is the depth of rainfall in inches for the rth increment of a storm hyetograph which is divided into m parts, each with essentially constant rainfall intensity (inch h^{-1}).

For each increment of the storm, e_r is determined using the following equation:

$$er = 1099\,[1-0.072^{(-.27} ir^{)}] \quad (8)$$

where i_r is the approximated constant intensity over the time increment. The value of v_r is calculated by the following:

$$v_r = i_r \times t_r \quad (9)$$

Other parts of the world may not have PIDF raster data to develop a R-factor or isoerodent map. Rainfall data is available in most developing countries but is obtained from non-recording-type rain gauges. This means that little quantitative information is available on the magnitude of instantaneous intensity (i.e. for 30 min) and the energy of a rain storm for a longer period. As a result, simple empirical formula for estimating the rainfall erosivity index (R-factor) from rainfall amounts is used. Petebendege (1990) has found a relationship between the 'mean annual erosivity', 'mean annual rainfall', and 'rainfall aggressive index' by multivariable regression analysis. The equation derived is as follows.

$$MAE = (5.62 \times MAR) + (92.32 \times RA) - 11754.49 \quad (10)$$

where MAE = mean annual erosivity (in J km^{-2})

MAR = mean annual rainfall (in mm)

RA = rainfall aggressiveness index

4 Introducing the *K*-factor (soil erodibility factor) into universal soil loss equation/revised universal soil loss equation/revised universal soil loss equation 2

Auerswald et al. (2014) have reviewed how the soil erodibility factor (*K*-factor) was developed by Wischmeier and Smith (1965a) and its subsequent development. In the UK, the *K*-factor as an attribute is available with gSSURGO data. However, setting up and collecting data from experimental runoff plots to establish the *K*-factor is expensive and time consuming. Many developing countries lack even simple soil data (Panda et al., 2004). In these cases, the *K*-factor can be developed using the Nomograph process that uses five factors of soil (calculated as percentages):

- Silt + very fine sand (0.002–0.10 mm);
- Sand (0.10–2.0 mm);
- Organic matter;
- Soil structure; and
- Permeability.

The soil erodibility factor '*K*' can be evaluated on experimental plots using the following equation:

$$K = A/\text{RLSCP} \text{ (for non-standard conditions)} \quad (11)$$

$K = A/R$ (for standard condition i.e. slope gradient = 9%, slope length = 22.1 m, cultivated continuous fallow land and ploughing up and down the slope. Here L = S = C = P = 1) (12)

5 Introducing the *L*-factor (slope length factor) and the *S*-factor (slope gradient factor) into universal soil loss equation/revised universal soil loss equation/revised universal soil loss equation 2

Slope length is defined as 'the distance from the point of origin of over land flow to either:'

- The point where the slope decreases to the extent that depositions begins; and
- The point where the run off enters a well-defined channel that may be a part of drainage network or a constructed channel.

The slope length factor (*L*) is defined as:

$$L = (\lambda/22.1)m \quad (13)$$

where λ = field slope length (in m)

m = exponent influenced by the interaction of slope length with gradient and may be influenced by the soil properties and the type of vegetation.

= 0.3 (for long slopes with gradient <0.5%)

= 0.6 (for slope gradient >10%)

= 0.5 (the mostly acceptable average value)

The United States Soil Conservation Service (US-SCS) (1972) developed an equation modifying the slope gradient parameter determining formula of Wischmeier (1959). The equation in its modified form is given as:

$$S = (0.43 + 0.30\,s + 0.043\,s^2)/6.613 \tag{14}$$

where s = slope gradient (in %)

This slope gradient equation was modified by Wischmeier and Smith (1978) to a form of slope gradient related to the slope steepness in 'degree' form. The equation is as illustrated in Equation 15. This equation was established using numerous test plot experiments:

$$S = 65.41\sin^2\theta + 4.56\sin\theta + 0.065 \tag{15}$$

where S = slope gradient factor and θ = angle of slope in degrees

The US-SCS (1972) developed an equation to account for the combined effect of slope length and slope gradient on soil loss. The equation holds good for slope gradient up to 20% and field slope length of up to 350 m:

$$LS = \sqrt{\lambda\left(0.0138 + 0.00965s + 0.00138s^2\right)} \tag{16}$$

where λ = measured slope length in the field, s = slope gradient in percentage

The equation for 'LS'-factor calculation for slope gradient from 10% to 50% and slope length up to 800 m is as follows:

$$LS = (\lambda / 22.1)0.6\,(s / 9)1.4 \tag{17}$$

where λ = measured slope length in the field, s = slope gradient in percentage

The US-SCS (1972) has developed a methodology for calculating the combined 'LS'-factor for all kinds of slopes and slope lengths. The equation is represented as:

$$LS = \sqrt{\lambda / 100\left(0.136 + 0.097s + 0.0139s^2\right)} \quad 18$$

where λ = measured slope length in the field, s = slope gradient in percentage

A newly modified equation combining 'L-' and 'S'-factor can be used for all conditions of slope length and gradient:

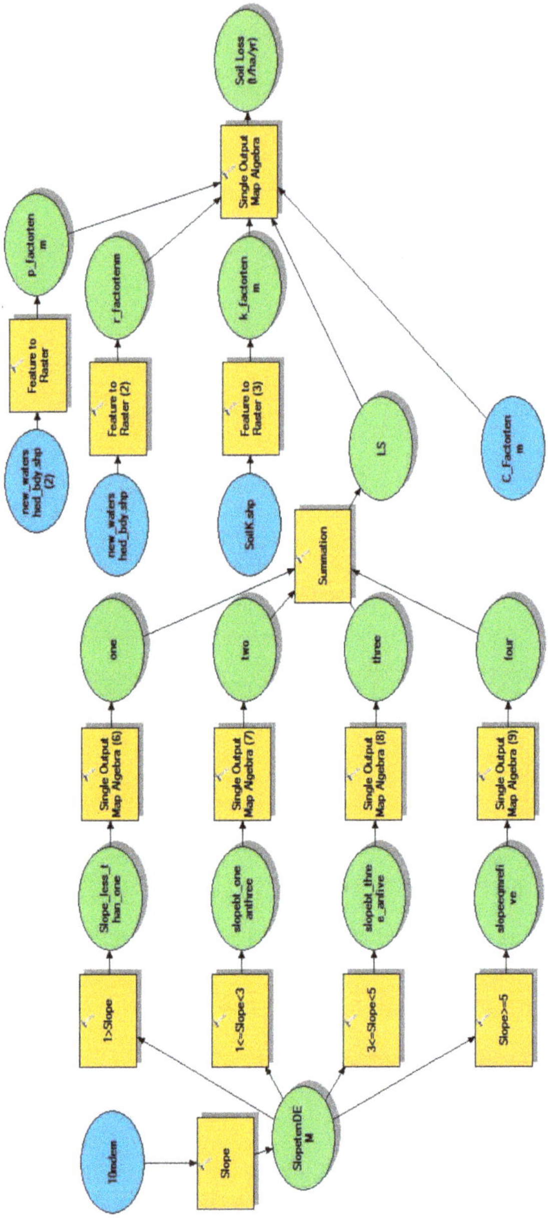

Figure 1 Automated geospatial model developed to calculate annual soil erosion load to Lake Nickerboker (Ada Creek retention pond), an example.

$$LS = (\lambda/22.13)m\,[65.41\sin^2\theta + 4.56\sin\theta + 0.065] \tag{19}$$

where λ = measured slope length in the field, θ = angle of slope in degrees, m = exponent factor varying from 0.2 to 0.5 and mostly 0.5 is used.

More recently the *LS*-factor has been calculated using the formula:

$$\mathrm{LS} = [0.65 + 0.456\,(\mathrm{slope}) + 0.00654\,(\mathrm{slope})^2\,]\,(98.4/72.5)^{NN} \tag{20}$$

where NN is 0.2 for slope less than 1%, 0.3 if slope is between 1% and 3%, 0.4 if slope is between 3% and 5%, or 0.5 for slopes greater than 5% that can be implemented in ArcGIS ModelBuilder (Panda et al., 2011).

Individual slope raster data can be created by selecting cell values of less than 1%, between 1% and 3%, between 3% and 5%, and slopes greater than 5% using the classified slope raster of the study area as shown in Fig. 1. Each slope raster gets transformed into the individual LS raster through the use of 'Map Algebra' tool, which can multiply the raster cell values with the LS equation shown in Equation 20. For each slope raster, the NN value is changed according to the corresponding values of 0.2, 0.3, 0.4, or 0.5. Once all four LS rasters are created, they are summed (spatially) together using the 'Sum' tool available with the Math Toolset of Spatial Analyst Toolbox of ArcToolbox (Fig. 1). Geospatial modelbuilders can obtain the final *LS*-factor raster using this method in ArcGIS. Separate *L*- and *S*-factors are also available separately as attributes for most parts of USA via gSSURGO or STASGO2 data. USDA developed gSSURGO (10m spatial resolution Gridded SSURGO raster) data contains L- (Slope Length - USLE) and S- (Slope Gradient - USLE) factors to be directly embedded to the Modified RUSLE model. However, it is not available for other parts of the world. Researchers should rely on the above L-, S-, or LS-factor development algorithms presented above.

6 Introducing the *C*-factor (cropping/cover management factor) and the *P*-factor (area control practice factor) into universal soil loss equation/revised universal soil loss equation/revised universal soil loss equation 2

The cropping/cover management (*C*) factor describes the total effect of vegetation, residue, soil surface, and management on soil loss. The value is typically not constant over the year. Although treated as an independent variable in the equation, the true value of this factor is potentially dependent upon all other factors. Hence the value of the '*C*'-factor needs to be established experimentally in local conditions. Land cover classified maps help to obtain a spatial *C*-factor using *C*-factor tables (database) developed by a wide range of researchers (see review in Gabriels et al., 2003). Panda et al. (2022) have developed a *C*-factor in ArcGIS ModelBuilder using the NDVI-based formula

($C = 0.1\left(\frac{NDVI+1}{2}\right)$)), and a *P*-factor using the classified land-use map of the study area with the combination of NDVI classification and land conservation spatial data available for the study area. It is to be noted that many country's (including United States) catalog all conservation lands as GIS vector layers and they can be used for P-factor raster development along with vegetation density information obtained from NDVI classification raster.

7 Case study: using high spatial resolution precipitation data for *R*-factor estimation

Rasterized rainfall data, such as PRISM data from a 4-km grid, were first introduced in *R*-factor calculation for RUSLE2 in the 1990s (Daly et al., 1997). *R*-factor values can be computed accurately on a spatial (grid) basis with the latest weather based I_{30} data availability in <1 km^2 (~800 m × 800 m) grid format via the NOAA (http://hdsc.nws.noaa.gov/hdsc/pfds/pfds_gis.html) (Bonnin et al., 2006). This improvement represents a leap forward in estimation. With these data, Equations 4 and 5 can easily be replicated with the grid-based (raster) I_{30} and I data to produce timely *R*-factor for regions of USA. The *R*-factor calculation is potentially more accurate given the availability of recent precipitation data. Developing the *R*-factor and subsequent isoerodent map development could be facilitated by geospatial technology-based model development since precipitation records are available in raster format.

7.1. National Oceanic and Atmospheric Administration precipitation raster data format and description

The data used in this study (the State of Georgia as the study area, Fig. 2) for the *R*-factor calculation are the same 100-year, 30-min storm event raster data developed by NOAA, NWS HDSC which makes *R*-factor development more efficient with independent *R*-values calculated for each 0.64 km^2 grid or pixel. Metadata for storm event (I_{30}) data suggests that these data are extremely efficient for developing the *R*-factor for the US conditions The storm event raster was developed using precipitation records from the 250+ National Climatic Data Center (NCDC) stations collected at 15-min intervals, six Georgia Forestry Commission (GFC), Fire Weather System (FWS) stations at hourly intervals, four Remote Automatic Weather Stations (RAWS) stations at hourly intervals, two Fort and Voluntary Observers Database (FORTS) stations at hourly intervals, and one USGS station at hourly intervals (Perica et al., 2013).

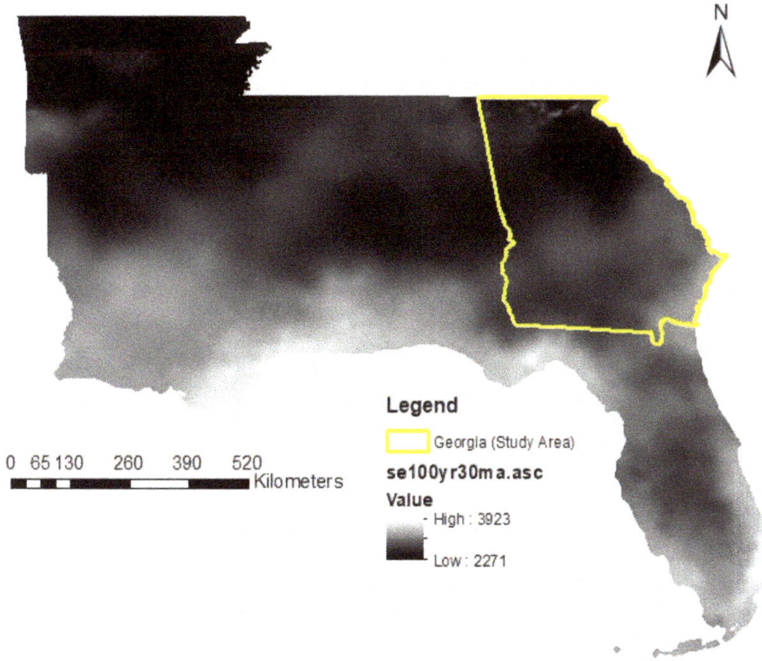

Figure 2 Partial duration series 100-year, 30-min precipitation frequency raster of southeastern USA. The study which is within the state boundary for Georgia, USA is outlined in yellow. Source: Adapted from: Datahttps://www.nws.noaa.gov/oh/hdsc/PF_documents/Atlas14_Volume9.pdf .

7.2. Isoerodent map development using National Oceanic and Atmospheric Administration precipitation raster

This study for estimating the *R*-factor covered the State of Georgia in the United States of America (USA). The raster file (se100yr30ma.asc) covering southeastern US (Fig. 2) was downloaded from https://www.nws.noaa.gov/oh/hdsc/PF_documents/Atlas14_Volume9.pdf (NOAA, 2019). This represents a partial duration series precipitation frequency analysis of 100-year average recurrence interval with 30-min duration data. The raster data is of 0.64 km^2 spatial resolution and according to NOAA (2019), *'these precipitation frequency (PF) estimates are used as basic design criteria for a wide variety of hydraulic structures such as culverts, roadway drainages, bridges and small dams. This dataset is for the display and/or analyses of spatially distributed point PF estimates'*. The weather data used for the I_{30} raster development ranges from as early as second decade of the twentieth century until the latest in 2011 (Perica et al., 2013). Annual average precipitation data (1980–2010) for Georgia, USA

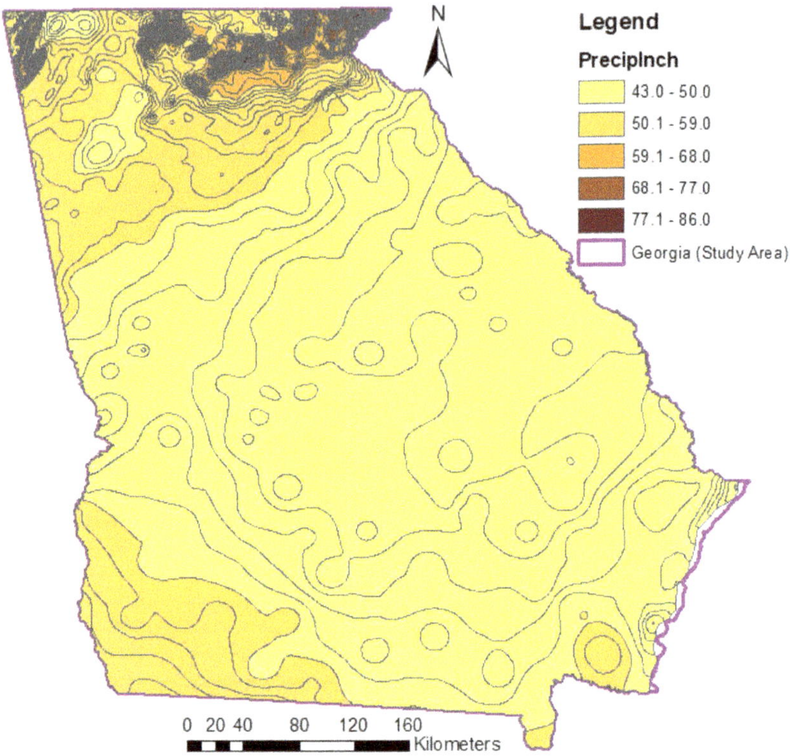

Figure 3 Annual average precipitation distribution vector of Georgia. Source: Adapted from: https://datagateway.nrcs.usda.gov/.

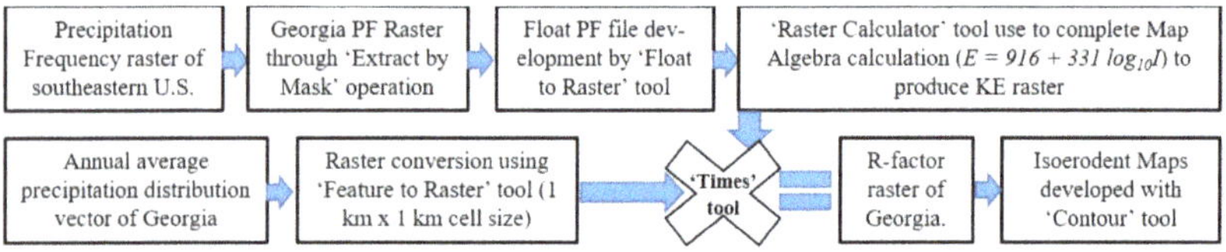

Figure 4 Basic procedural flow chart for *R*-factor and isoerodent map development for Georgia using NOAA PF raster and the average annual precipitation distribution vector.

was downloaded in vector format (Fig. 3) from the USDA NRCS Geospatial Data gateway (https://datagateway.nrcs.usda.gov/). The data was available at the same 0.64 km^2 spatial resolution grid (raster) using the 'precipitation in inch' attribute to suit to the spatial resolution of NOAA PF data for Georgia, USA. Step-by-step processes (Fig. 4) were followed as directed by the web interface to obtain the *R*-factor raster and subsequent isoerodent map of different unit intervals for Georgia.

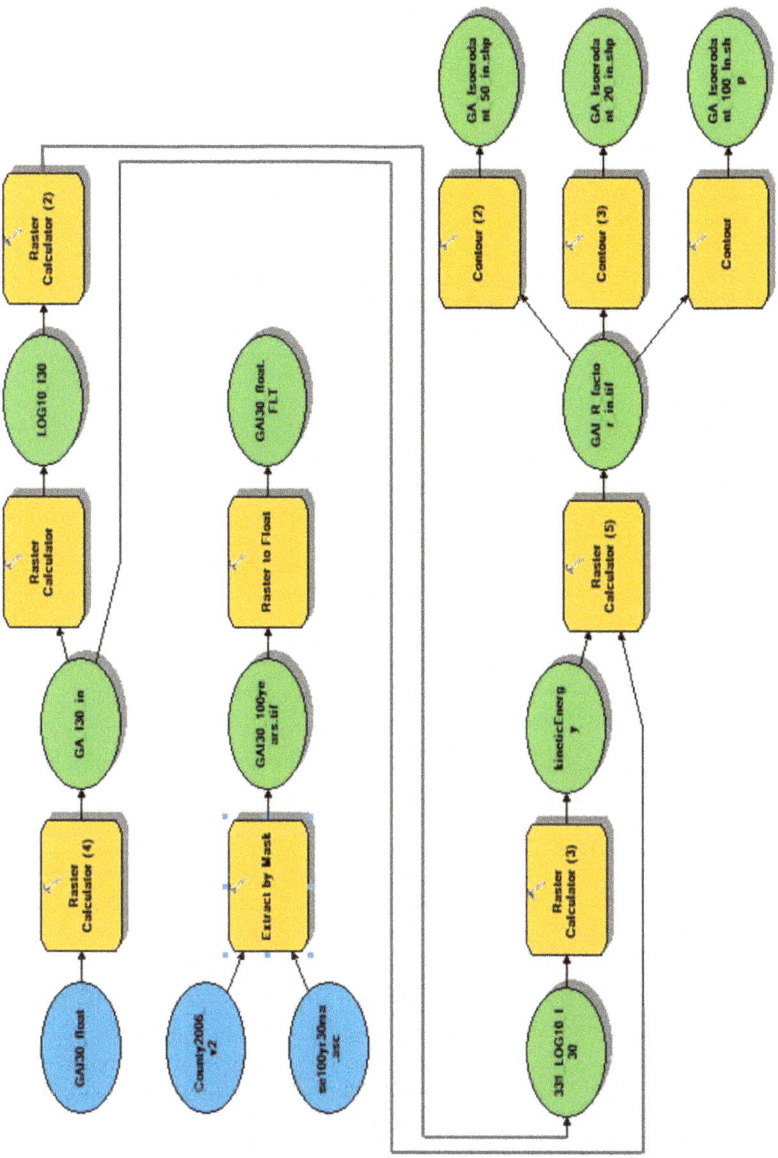

Figure 5 Automated geospatial model framework for obtaining *R*-factor raster and isoerodent maps vector of Georgia.

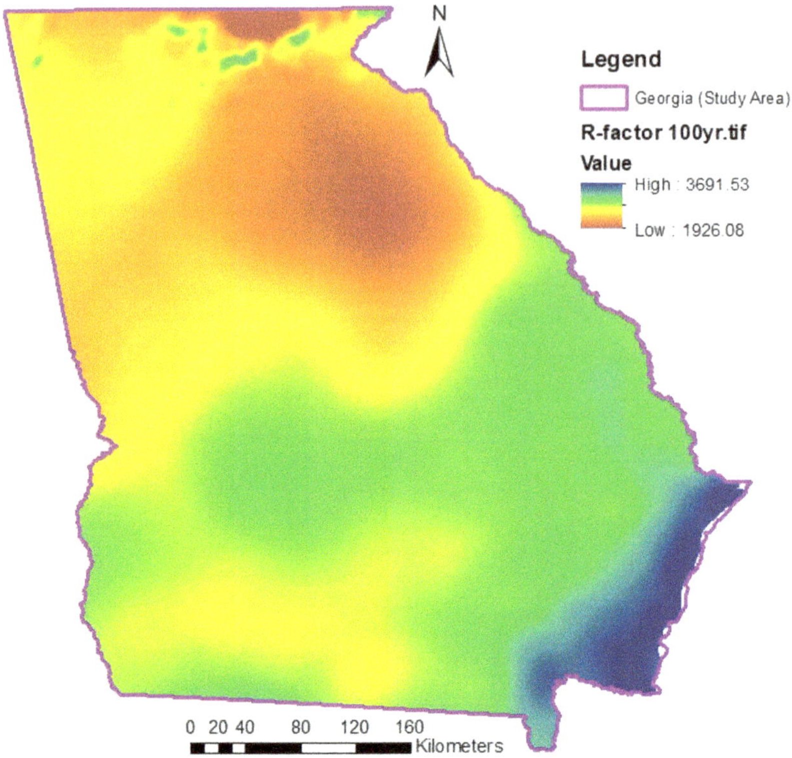

Figure 6 *R*-factor map of Georgia developed with latest NOAA developed precipitation frequency.

A Georgia USA shapefile was used as mask with the se100 year 30-min raster data to produce the study area [the area of interest (AOI)] PF raster which was subsequently converted to a float file to facilitate raster calculation (Map Algebra) with ArcGIS 10.5 software (Fig. 5). This float file served as the primary input for the rest of the automated geospatial model developed in the ArcGIS ModelBuilder platform. The input float file was used as input for *I* in the equation '$E = 916 + 331 \log_{10} I$' in ArcGIS 'Raster Calculator' tool with proper algebraic function input. In order to make the overall raster calculation process smoother and efficient, a series of rasters, which included $\log_{10} I$, $331 \log_{10} I$, and $916 + 331 \log_{10} I$ rasters, were developed. The final raster developed was the kinetic energy (*E*) raster. The *E* raster was multiplied by the Georgia, USA, annual average precipitation raster to account for Equation 5 of Wischmeier and Smith (1965a,b). We were thus able to develop a new Georgia USA *R*-factor raster (Fig. 6).

Isoerodent maps of Georgia, USA, at various intervals, i.e. 100 m (Fig. 7a), 50 m (Fig. 7b), and 20 m (Fig. 7c), respectively, were developed using the contour tools described earlier. This makes it possible to generate erosion index values

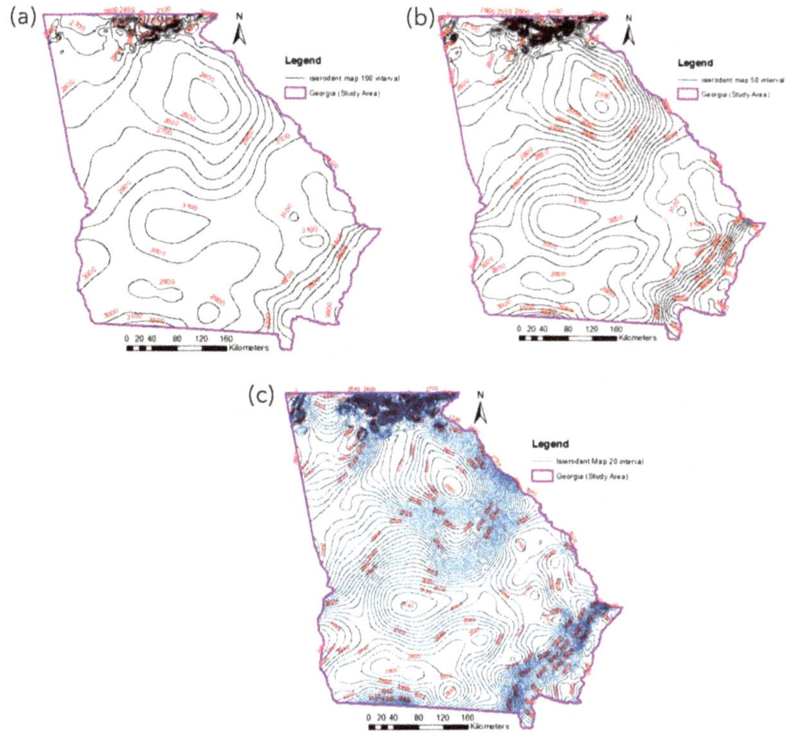

Figure 7 Isoerodent map of Georgia at 100 (a), 50 (b), and 20 (c) unit interval developed with latest NOAA developed precipitation frequency data.

for a smaller spatial extent for soil erosion quantification. However, the *R*-factor will then be unique for a ~0.7 km^2 grid or pixel based on the NOAA spatial rasters used in this study. Finally, the isoerodent results from our study were compared with the old isoerodent map of Georgia, USA (Fig. 8). The *R*-factor calculated with latest climate NOAA raster data produced high erosion index (*R*-factor) values, i.e. ~six to seven times higher than the earlier isoerodent map for Georgia.

8 Conclusion and future trends

Individual geospatial models for *K*-, *L*-, and S-factors are being developed using gSSURGO soil data. *C*- and *P*-factors are being developed in the ArcGIS ModelBuilder using satellite remote sensing data and a normalized difference vegetative index (NDVI) approach with updated algorithms that comply with the latest changes in land use and conservation practices.

In cold climates, where frost forms during the winter season in the soil and where areas are also experiencing increasing amounts of snow precipitation due to climate change, the RUSLE model needs to be modified significantly in

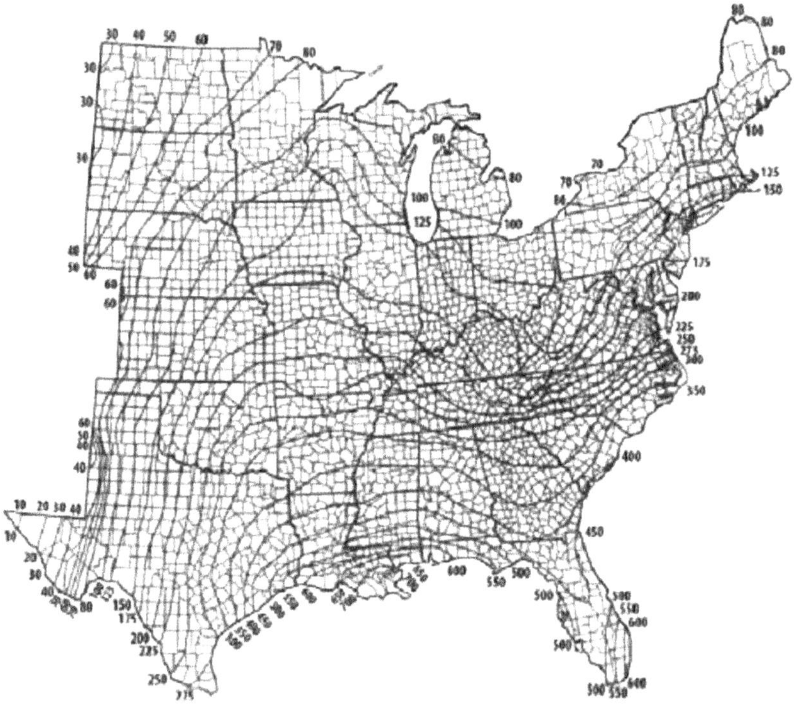

Figure 8 Current isoerodent map of eastern USA (EPA, 2001).

Figure 9 Photograph of a typical thermokarst from the Caribou-Poker Creek Watershed of Alaska.

order to quantify the amount of erosion or erodibility of these areas. In Arctic and the sub-Arctic regions, such as Alaska, the soil is prone to erodibility due to snowmelt and/or melting of permafrost-forming thermokarsts (see Fig. 9; Misra et al., 2011). Thermokarsts are formed when the permafrost melts and induces thaw slump. The loose slumped or collapsed soil is then eroded due to surface rainfall, subsequent snowmelt or groundwater pressure. The snowmelt erosion is only possible when the shear force exerted by the snowmelt exceeds the soil matrix strength bounded by the soil frost (Daanen et al., 2011). This is also true for soil erosion in Wisconsin or other northern cold climate regions of the USA.

As noted earlier, the new *R*-factor values established through the case study differ from the old isoerodent map of Georgia. NOAA uses current precipitation data (through to 2013 and many spatial cases through 2019) which better reflect recent shifts in precipitation patterns. Since they were not able to do this, the authors recognize the need for verification of the *R*-factors developed by this procedure with field experimentation. The goal of the case study was to develop an automated model for efficiently developing latest *R*-factors of any spatial locations using PF rasters created by NOAA. The authors are applying this technique in three large watersheds in a forthcoming publication on 'streambank erosion vulnerability assessment (SBEVA) and RUSLE modeling supported forest structure vulnerability analysis and decision support.' This paper has been published in the journal *Environmental Modeling and Software* (Panda et al., 2022).

9 Acknowledgement

The authors would like to acknowledge the USDA Forest Service Southern Research Station (SRS) and Region 8 for funding the Georgia-based study. Furthermore, the authors extend thanks to the reviewers and editorial board members for this manuscript's review and publication.

10 References

Almagro, A., Oliveira, P. T. S., Nearing, M. A. and Hagemann, S. (2017). Projected climate change impacts in rainfall erosivity over Brazil. *Scientific Reports* 7(1), 8130.

Amatya, D. M., Sun, G. and Panda, S. S. (2017). Ecohydrological studies at Santee experimental forest, baseline data, extreme events, and hydrological vulnerabilities in the context of south Eastern US forests. Closing Keynote presented in the 2017 Municipal Wet Weather Stormwater Conference May 17, 2017, Charleston, SC.

Amatya, D. M., Tian, S., Marion, D. A., Caldwell, P., Laseter, S., Youssef, M. A., Grace, J. M.,Chescheir, G. M., Panda, S. S., Ouyang, Y., Sun, G. and Vose, J. M. (2021). Estimates of precipitation IDF curves and design discharges for road-crossing drainage structures: case study in four small forested watersheds in the Southeastern US.

Journal of Hydrologic Engineering 26(4), 05021004. DOI: 10.1061/(ASCE)HE.1943-5584.0002052. © 2021 American Society of Civil Engineers.

Amatya, D. M., Williams, T. M., Nettles, J. E., Skaggs, R. W. and Trettin, C. C. (2019). Comparison of hydrology of two Atlantic coastal plain forests. *Transactions of the ASABE* 62(6), 1509–1529.

Auerswald, K., Fiener, P., Martin, W. and Elhaus, D. (2014). Use and misuse of the K factor equation in soil erosion modeling: an alternative equation for determining USLE nomograph soil erodibility values. *CATENA* 118, 220–225.

Biasutti, M. and Seager, R. (2015). Projected changes in US rainfall erosivity. *Hydrology and Earth System Sciences* 19(6), 2945–2961.

Bonnin, G. M., Martin, D., Lin, B., Parzybok, T., Yekta, M. and Riley, D. (2006). Precipitation-frequency atlas of the United States. *NOAA Atlas* 14(2), 1-65.

Browning, G. M., Parish, C. L. and Glass, J. (1947). A method for determining the use of limitations of rotation and conservation practices in the control of soil erosion in Iowa. *Journal of the American Society of Agronomy*. 39(1), 65-73.

Daanen, R. P., Misra, D. and Thompson, A. (2011). Frozen soil hydrology. In: Haritashya, U. K., Singh, V. and Singh, P. (Eds) *Encyclopedia of Snow, Ice, and Glaciers*. Springer Press, USA. Part 6, 306–311.

Daggupati, P., Sheshukov, A., Douglas-Mankin, K. R., Barnes, P. L. and Delvin, D. L. (2009). Field-scale targeting of cropland sediment yield using ArcSWAT. In: Proceedings of the 5th International SWAT Conference, Boulder, CO, August 5–7, 2009.

Daly, C., Taylor, G. and Gibson, W. (1997). The PRISM approach to mapping precipitation and temperature. In: Proceedings of the 10th Conference on Applied Climatology, Reno, NV, October 20–24, 1997. Am. Table 4 continued from previous page.Meteorological Society, Boston, MA, pp. 20–23.

Feaster, T. D., Gotvald, A. J. and Weaver, J. C. (2014). *Methods for Estimating the Magnitude and Frequency of Floods for Urban and Small, Rural Streams in Georgia, South Carolina, and North Carolina, 2011*. United States Geological Survey, Columbia.

Gabriels, D., Ghekiere, G., Schiettecatte, W. and Rottiers, I. (2003). Assessment of USLE cover-management C-factors for 40 crop rotation systems on arable farms in the Kemmelbeek watershed, Belgium. *Soil and Tillage Research* 74(1), 47–53.

Giang, P. Q., Giang, L. T. and Toshiki, K. (2017). Spatial and temporal responses of soil erosion to climate change impacts in a transnational watershed in Southeast Asia. *Climate* 5(1), 22.

Hosking, J. R. M. and Wallis, J. R. (2005). *Regional Frequency Analysis: An Approach Based on L-Moments*. Cambridge University Press, UK.

Hrachowitz, M., Savenije, H., Blöschl, G., McDonnell, J., Sivapalan, M., Pomeroy, J. W., Arheimer, B., Blume, T., Clark, M. P., Ehret, U., Fenicia, F., Freer, J. E., Gelfan, A., Gupta, H. V., Hughes, D. A., Hut, R. W., Montanari, A., Pande, S., Tetzlaff, D., Troch, P. A., Uhlenbrook, S., Wagener, T., Winsemius, H. C., Woods, R. A., Zehe, E. and Cudennec, C. (2013). A decade of Predictions in Ungauged Basins (PUB)–a review. *Hydrological Sciences Journal* 58(6), 1198–1255.

Kuichling, E. (1889). The relation between the rainfall and the discharge of sewers in populous districts. *Transactions of the American Society of Civil Engineers* 20(1), 1-56.

Laflen, J. M. and Flanagan, D. C. (2013). The development of US soil erosion prediction and modeling. *International Soil and Water Conservation Research* 1(2), 1-11.

Laseter, S. H., Ford, C. R., Vose, J. M. and Swift, L. W. (2012). Long-term temperature and precipitation trends at the Coweeta Hydrologic Laboratory, Otto, NC, USA. *Hydrology Research* 43(6), 890-901.

Licciardello, F., Ciampalini, R., Pastor, A., Huard, F., Follain, S., Crabit, A., Le Bissonnais, Y. and Raclot, D. (2019, January). Modelling the impact of climate change and land use on soil erosion in a Mediterranean catchment. *Geophysical Research Abstracts* 21, P1-1, 1.

Lufafa, A., Tenywa, M. M., Isabirye, M., Majaliwa, M. J. G. and Woomer, P. L. (2003). Prediction of soil erosion in a Lake Victoria basin catchment using a GIS-based universal soil loss model. *Agricultural Systems* 76(3), 883-894.

Millward, A. A. and Mersey, J. E. (1999). Adapting the RUSLE to model soil erosion potential in a mountainous tropical watershed. *CATENA* 38(2), 109-129.

Misra, D., Daanen, R. P. and Thompson, A. (2011). Thermokarst. In: Haritashya, U. K., Singh, V. and Singh, P. (Eds) *Encyclopedia of Snow, Ice, and Glaciers*. Springer Press, USA. Part 19, 1158-1166.

Moran, M. S., Peters, D. P. C., McClaran, M. P., Nichols, M. H. and Adams, M. B. (2008). Long-term data collection at USDA experimental sites for studies of ecohydrology. *Ecohydrology* 1(4), 377-393.

Moreiras, S. M., Pont, I. V. D. and Araneo, D. (2018). Were merely storm-landslides driven by the 2015-2016 Niño in the Mendoza River valley? *Landslides* 15(5), 997-1014.

Musgrave, G. W. (1947). The quantitative evaluation of factors in water erosion: a first approximation. *Journal of Soil and Water Conservation* 2, 133-138.

Nearing, M. A. (2001). Potential changes in rainfall erosivity in the United States with climate change during the 21st century. *Journal of Soil and Water Conservation* 56, 229-232.

NOAA (2019). NOAA atlas 14 precipitation frequency estimates in GIS compatible format. Available at: https://hdsc.nws.noaa.gov/hdsc/pfds/pfds_gis.html. Alabama, Georgia Isopluvials of 1-Year 60-Minute Precipitation in Inches NOAA Atlas 14 Volume 9, Version 2. *Southeastern States* (2nd edn., Volume 14), National Oceanic and Atmospheric Administration, United States, 2013. Accessed on April 16, 2019.

NRCS (2004). *National Engineering Handbook: Part 630–Hydrology*, United States Department of Agriculture Soil Conservation Service, Washington, DC.

ODOT (2014). Roadway drainage manual, Chapter 7. In: *Hydrology*. Oklahoma Department of Transportation. Available at: https://www.ok.gov/odot/documents/Chapter%207%20Hydrology.pdf. Accessed on August 4, 2022.

Ouyang, D. and Bartholic, J. (2001). Web-based GIS application for soil erosion prediction. In: Proceedings of an International Symposium–Soil Erosion Research for the 21st Century Honolulu, HI, January 3-5.

Panagopoulos, T. and Ferreira, V. 2010, July. Decision support system for erosion risk assessment. In: Proceedings of the 14th WSEAS International Conference on Systems: Part of the 14th WSEAS CSCC Multiconference (Volume I, pp.76-80). World Scientific and Engineering Academy and Society (WSEAS).

Panda, S. S., Amatya, D., Grace, J. M., Caldwell, P. and Marion, D. (2018). Automated geospatial model based assessment of erosion vulnerability at forest road/ Stream crossings under extreme precipitation intensities scenario. Accepted to be published in the Proceedings and be presented in the Sixth Interagency Conference on Research in the Watersheds (ICRW) at National Conservation Training Center, Shepherdstown, WV, July 23-26, 2018.

Panda, S. S.,Amatya, D., Grace, J. M., Caldwell, P. and Marion, D. (2019). Forest road/stream crossings erosion vulnerability assessment using geospatial hydrologic model under extreme precipitation events. Presented in the 2019 American Geophysical Union (AGU) Conference, San Francisco, CA, December 9–13, 2019.

Panda, S. S., Amatya, D. M., Grace, J. M., Caldwell, P. and Marion, D. A. (2022). Extreme precipitation-based vulnerability assessment of road-crossing drainage structures in forested watersheds using an integrated environmental modeling approach.*Environmental Modelling and Software*, 105413. DOI: 10.1016/j.envsoft.2022.105413.

Panda, S. S., Andrianasolo, H., Murty, V. V. N. and Nualchawee, K. (2004). Forest management planning for soil conservation using satellite images, GIS mapping, and soil erosion modeling. *Journal of Environmental Hydrology* 12(13), 1–16.

Parry, M. L., Canziani, O. F., Palutikof, J. P., van der Linden, P. J. and Hanson, C. E. (2007). *Climate Change 2007: Working Group II: Impacts, Adaptation and Vulnerability*. Cambridge University Press, Cambridge.

Perica, S., Martin, D., Pavlovic, S., Roy, I., St Laurent, M., Trypaluk, C., Unruh, D., Yekta, M. and Bonnin, G. (2013). *NOAA Atlas 14 Precipitation-Frequency Atlas of the United States*, National Oceanic and Atmospheric Administration, Silver Spring, MD.

Perica, S., Pavlovic, S., Laurent, M. S., Trypaluk, C., Unruh, D., Martin, D. and Wilhite, O. (2013). *NOAA Atlas 14: Precipitation Frequency Atlas of the United States Volume 10: Northeastern States*. NOAA, National Weather Service, Silver Spring, MD.

Pruski, F. F. and Nearing, M. A. (2002). Climate-induced changes in erosion during the 21st century for eight U.S. locations. *Water Resources Research* 38(12), 1298.

Renard, K. G., Foster, G. R., Weesies, G. A., McCool, D. K. and Yoder, D. C. (1997). Predicting rainfall erosion by water: a guide to conservation planning with the revised universal soil loss equation (RUSLE). In: *Agriculture Handbook*, No. 703. US Government Printing Office, Washington, DC.

Renard, K. G., Foster, G. R., Weesies, G. A. and Porter, J. P. (1991). RUSLE: revised universal soil loss equation. *Journal of Soil and Water Conservation* 46(1), 30–33.

Rogger, M., Kohl, B., Pirkl, H., Viglione, A., Komma, J., Kirnbauer, R., Merz, R. and Blöschl, G. (2012). Runoff models and flood frequency statistics for design flood estimation in Austria–do they tell a consistent story? *Journal of Hydrology* 456, 30–43.

Schwab, G. O., Frevert, R. K., Barnes, K. K. and Edminster, T. W. (1971). *Elementary Soil and Water Engineering*, John Wiley & Sons, Inc., New York, 316 p.

Sharma, J. B. (2008). Erosion and sediment modeling of the Lake Lanier watershed. Ph.D. Thesis, Department of Geography, University of Georgia, Athens, GA, p. 187.

Smith, D. D. (1941). Interpretation of soil conservation data for field use. *Agricultural Engineering* 22, 173–175.

USDA-ARS (2013). *Science Documentation: Revised Universal Soil Loss Equation, Version 2*. USDA-ARS, Washington, DC.

US Water Resource Council (1981). Estimating peak flow frequencies for natural ungaged watersheds.

Van Remortel, R. D., Maichle, R. W. and Hickey, R. J. (2004). Computing the LS Factor for the revised universal soil loss equation through array-based slope processing of digital elevation data using C++ executable. *Computers and Geosciences* 30(9–10), 1043–1053.

Wagner, D. M., Krieger, J. D. and Veilleux, A. G. (2016). Methods for estimating annual exceedance probability discharges for streams in Arkansas, based on data through water year 2013. *Scientific Investigations Report 2016-5081*. U.S. Geological Survey, Reston, VA, 149 p.

Wischmeier, W. H. (1959). A rainfall erosion index for a universal soil-loss equation. *Soil Science Society of America Journal* 23(3), 246-249.

Yitayew, M., Pokrzywka, S. J. and Renard, K. G. (1999). Using GIS for facilitating erosion estimation. *Applied Engineering in Agriculture* 15(4), 295-301.

Zhang, G. H., Nearing, M. A. and Liu, B. Y. (2005). Potential effects of climate change on rainfall erosivity in the Yellow River Basin of China. *Transactions of the American Society of Agricultural Engineers* 48, 511-517.

Zhang, X. C. and Nearing, M. A. (2005). Impact of climate change on soil erosion, runoff, and wheat productivity in Central Oklahoma. *CATENA* 61(2-3), 185-195.

Zingg, A. W. (1940). Degree and length of land slope as it affects soil loss in runoff. *Agriculturists* 21, 59-64.

Part 3

Mitigating soil erosion

Chapter 7

Assessing the effectiveness and history of buffer strip implementation in preventing soil sediment and nutrient losses to surface waters

Brian Kronvang and Dominik Zak, Aarhus University, Denmark; Eva Skarbøvik and Anne-Grete Buseth Blankenberg, Norwegian Institute of Bioeconomy Research (NIBIO), Norway; and Marc Stutter, The James Hutton Institute, UK*

1 Introduction

Buffer strips along streams and around lakes are one of the most popular and commonly adopted measures in the landscape to mitigate the adverse impacts on surface waters from sediment and associated nutrient losses due to soil erosion and surface runoff from farmland (Haddaway et al., 2018; Stutter et al., 2012, 2018). Narrow strips of unmanaged and unfertilised land situated between fields and surface waters can vary in width, soil characteristics and slope, as well as having different vegetation and management requirements, such as the need for cutting and harvesting of vegetation within the strip (Fig. 1). Buffer strips can be established as 'dry buffers', meaning no direct changes are imposed upon the way that water passes under the narrow strip of land. They can also be established and modified with respect to their subsurface hydrology as 'integrated buffer zones' (IBZs) that capture the tile drain flow

* Contact author: BKR@ECOS.AU.DK.

http://dx.doi.org/10.19103/AS.2023.0131.14

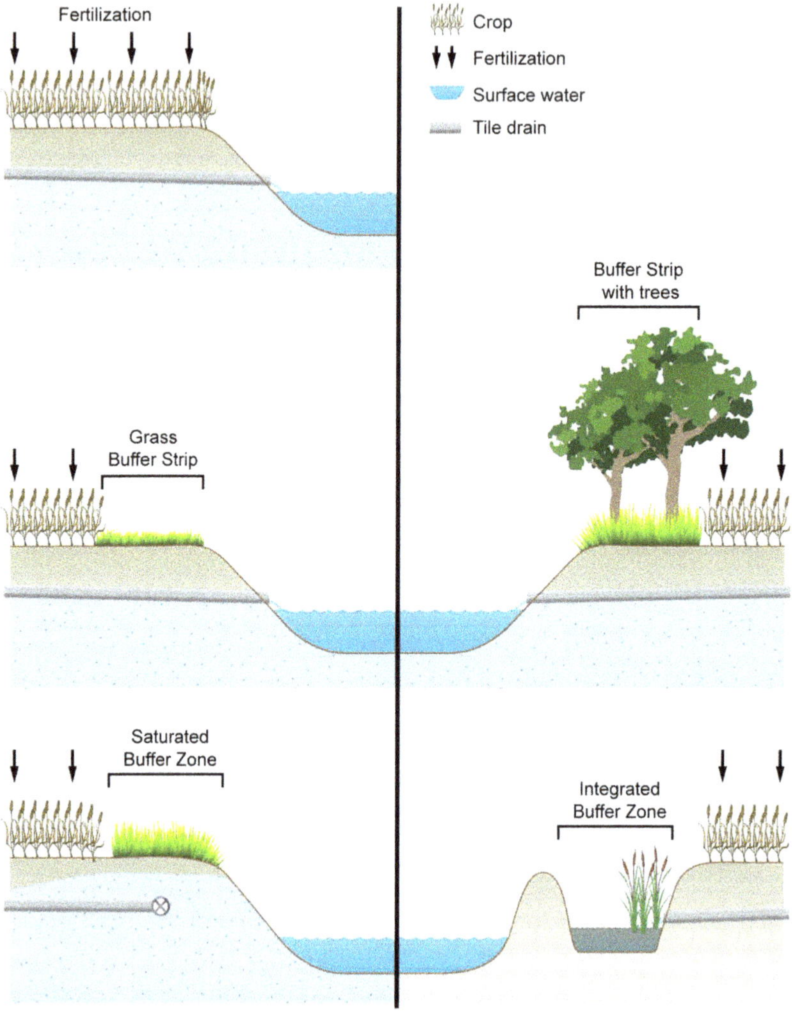

Figure 1 Different types of buffer strips varying in width and vegetation (grass and trees) with free drainage underneath (upper panel) and wetness as shown in the lower panel for a Saturated Buffer Zone (SBZ) and an Integrated Buffer Zone (IBZ), treating drainage water entering the buffer zone.

from the neighbouring fields (Zak et al., 2018, 2019). The distinctions between these two types of buffer strips are shown schematically in Fig. 1.

Buffer strips are one of the most commonly utilised nutrient management measures worldwide (Haddaway et al., 2018) and are principally designed and implemented to control sediment and phosphorus (P) losses to surface waters (Dillaha et al., 1988; Lee et al., 2000; Borin et al., 2005; Dorioz et al., 2006; Liu et al., 2009; Stutter et al., 2012, 2018, 2021a; Uusi-Kämppä et al., 2012; Gericke et al., 2020). The presence of densely rooted grass and herb vegetation, along with typically gentler slopes than in the adjoining fields, serves to decelerate uniform and concentrated surface flow and increase water infiltration. This reduction in flow velocity facilitates the deposition of sediment in field surface runoff, which decreases the transport of sediment and associated nutrients and other contaminants to surface waters (Dillaha et al., 1988; Daniels and Gilliam, 1996; Stutter et al., 2021a).

Vegetated buffer strips alongside stream channels are often left unmanaged and allowed to revert to their natural vegetation of grasses and herbs (Haddaway et al., 2018). However, buffer strips can also be managed by harvesting their biomass for use as fodder or biogas production (Moset et al., 2019; Stutter et al., 2019; Cole et al., 2020) or rewetting with tile drainage water, as is the case in saturated buffer zones (Jaynes and Isenhart, 2019) or integrated buffer zones (Zak et al., 2018, 2019). Another management measure includes planting trees that assist in reducing bank erosion and providing other important ecosystem services, such as producing wood and providing shade, protecting the water from extremes in temperature, etc. (Kronvang et al., 2012; Blankenberg and Skarbøvik, 2020).

Historically, buffer strips between fields and surface waters have been used as a mitigation measure to reduce the losses of sediment and sediment-associated contaminants from fields to the water (e.g. Dorioz et al., 2006). However, in the last two decades, research has started to focus on the multi-functionality of buffer strips relative to their effect on sediment and nutrient capture, their importance for surface water and terrestrial biodiversity and the transfer of other contaminants such as pesticides and provision of biomass (Fig. 2) (Stutter et al., 2012, 2019; Haddaway et al., 2018).

2 Effects of buffer strips

During the last five decades, numerous experimental studies around the world have focused on quantifying the effect of buffer strips on the capture of sediment and sediment-associated nutrients and the importance of factors such as buffer strip width, vegetation and management (e.g. Lee et al., 2000; Abu-Zreig et al., 2003; Liu et al., 2009; Søvik et al., 2012; Haddaway et al., 2018; Hille et al., 2019; Stutter et al., 2012, 2019, 2021a).

Fewer studies have investigated the function of buffer strips as sinks of dissolved substances such as nitrate-nitrogen (NO_3^--N) and dissolved reactive phosphorus (DRP) (Stutter et al., 2021). Dissolved substances can be involved in several biochemical reactions during their transport through the soil profile in the buffer strip with infiltrating water, such as sorption reactions (DRP) and

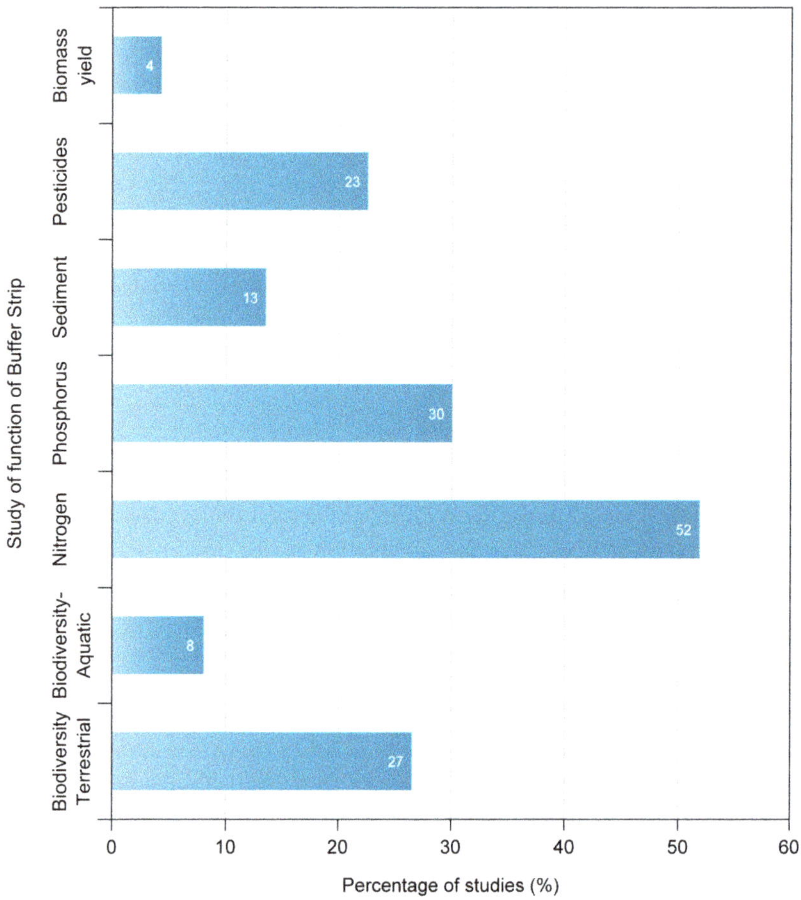

Figure 2 Percentage of studies on varying functions of buffer strips alongside streams in a mapping exercise based on international literature, including Europe, North America and Australia (*N* = 96 studies) conducted by Haddaway et al. (2018).

denitrification, and also in terms of plant uptake (NO_3^--N) (Hoffmann et al., 2009; Heinen et al., 2012; Hille et al., 2019; Stutter et al., 2021a).

2.1 Sediment

The effectiveness of buffer strips in terms of sediment capture has been demonstrated in numerous experimental studies worldwide (Haddaway et al., 2018; Stutter et al., 2021a). Several investigations have utilised published data to develop predictive statistical models - both simple regression models such as in Stutter et al. (2021a) as well as more sophisticated machine learning models (Gericke et al., 2020). Generally, the effectiveness of buffer strips in retaining sediment from soil erosion and surface runoff on adjacent fields

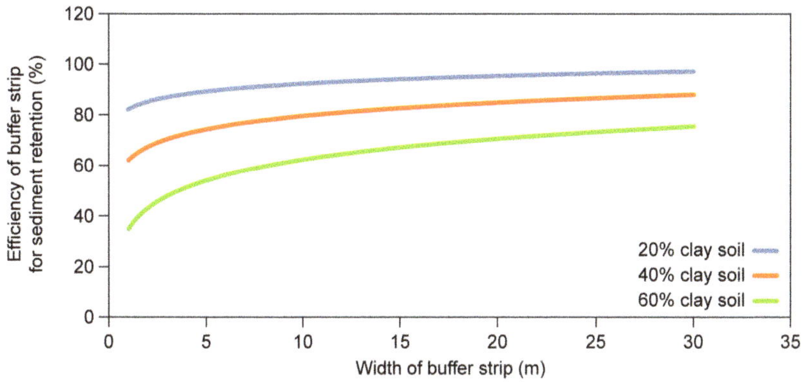

Figure 3 Effects of buffer strips of varying widths and percentages of clay soils on the retention of sediments using an equation from Stutter et al. (2021a), including data from 257 experimental studies.

depends on width, slope and soil type. In their comprehensive analysis of experimental data from 257 studies, Stutter et al. (2021a) found no significant effects on sediment retention of different vegetation types (grasses and herbs, trees, grass and trees, unmanaged set-aside) in the buffer strips. They also found that the average effectiveness of buffer strips in sediment retention amounted to approximately 72%, although this varied widely, from −36% to 100%. The width of the buffer strips was revealed to be much more important (13% in variance) than both soil type (4%) and slope (4%) (Stutter et al., 2021a) in terms of sediment trapping. The two-factor regression model developed by Stutter et al. (2021a) demonstrates the importance of width and soil type (clay content) in terms of the effectiveness of buffer strips in sediment retention (Fig. 3). The effect of buffer strips on retaining sediment from surface runoff thereby increases sharply with buffer strip width and decreases with the content of clay in buffer strip soils (Fig. 3).

2.2 Phosphorus

The effect of buffer strips on phosphorus retention has also been widely studied, albeit less extensively than the effect on sediment (Stutter et al., 2021a). Most studies have included the effects on total phosphorus (TP), whilst fewer have focused on the effect on dissolved phosphorus (DP) (Stutter et al., 2021a). The models based on the available data in international literature focus mainly on TP, and in the TP model developed by Stutter et al. (2021), width accounted for most of the variance in the dataset (N = 111; R^2 = 0.43). The median TP retention in the buffer strips in the Stutter et al. (2021a) dataset was 75%, ranging from −55% to 98%. As for sediment, the two factors in the model that accounted for most of the variation in the dataset were buffer

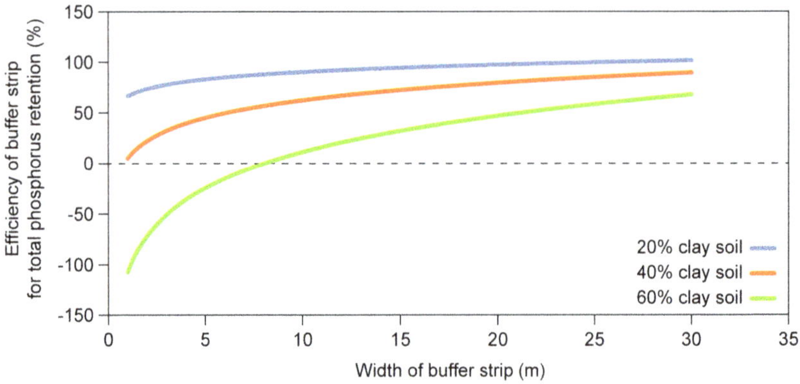

Figure 4 Effects of buffer strips with varying widths and clay soil percentages on the retention of total phosphorus, using the equation from Stutter et al. (2021a) based on data from 111 experimental studies.

strip width and clay percentage in the buffer strip soil (Fig. 4). The negative retention at low buffer strip widths (<10 m) and high soil clay content (>20%) shows the risk of dissolved reactive P re-mobilisation within the buffer strip (Dorioz et al., 2006; Stutter et al., 2009; Hoffmann et al., 2009). Hille et al. (2019) further investigated available management options to reduce P leaching from vegetated buffer strips and found that harvest of buffer biomass would, with time, decrease the potential P release and leaching, making a single annual harvest the most economical.

2.3 Nitrogen

The effectiveness of buffer strips for retention of total nitrogen (TN) and NO_3^--N has been widely discussed as the mix of surface runoff, tile drainage flow and groundwater flow is difficult to capture fully in experimental studies. Perhaps, the most comprehensive study on the subject to date was performed by Heinen et al. (2012) and Noij et al. (2012), and captured all water passing through buffer strips in experimental studies in the Netherlands. They found no evidence of N retention in their study of five sites with 5 m unfertilised buffer strips along ditches due to water passing underground as deep drainage through the buffer strips to surface waters. Only a limited effect of reduced nitrate leaching in the unmanaged 5 m buffer strips was found at one of the sites (Heinen et al., 2012 and Noij et al., 2012). The authors concluded that the 5 m wide buffer strips were ineffective at mitigating N loads from lowland agriculture to surface waters but that a better result would be obtained when designing strips specifically for abating surface runoff from more undulating landscapes.

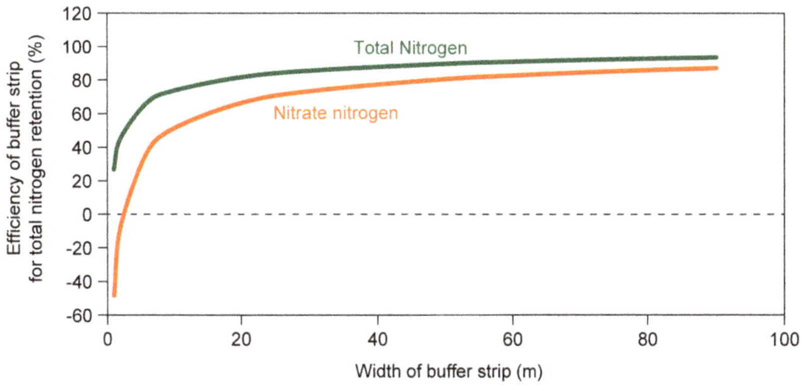

Figure 5 Effects of buffer strips of varying widths and clay soil percentages on the retention of total nitrogen and nitrate using equations from Stutter et al. (2021a), based on data from 40 and 94 experimental studies, respectively.

In a comprehensive review study by Stutter et al. (2021a), data on total N and NO_3^--N from 40 and 94 studies were extracted, respectively. Mean and median values for N retention (range) amounted to 32, 42 (−375% to 100%) and 6, 44 (-1650% to 100%), respectively. The huge range in values (including the negative retention values found for NO_3^--N in narrow buffer strips) highlights the challenges involved in accurately determining the effectiveness of buffer strips in terms of NO_3^--N retention. In the case of TN, retention was positive for all buffer strip widths, which might be due to the retention of particulate organic N from surface runoff.

A single factor model was developed for both nitrogen species in Stutter et al. (2021a), which showed a generally positive but weak explanatory power of the models (R^2 = 0.08 for TN; R^2 = 0.08 for NO_3^--N) (Fig. 5).

2.4 New methods for improving the effectiveness of buffer strips

Integrated buffer zones (IBZs) have recently been introduced in the Northwestern Europe temperate zone to improve delivery of ecosystem services compared with long-established vegetated buffer zones (Zak et al., 2019). The new IBZs tested in Denmark proved to be effective at intercepting tile drainage water and retaining NO_3^--N and DRP, as well as nutrients bound to particles (Zak et al., 2018, 2019). Specifically, in contrast to the traditional vegetated buffer strip, IBZs collect tile drainage water from the edge of the agricultural fields by combining a ditch-like pond, in which soil particles from both tile drain water and surface runoff can settle, and a flow-through filter bed (Fig. 1).

To improve nutrient removal in the IBZ, the filter bed can be planted with trees such as alder or willow that tolerate water-saturated soils for prolonged periods, such as in winter. It is important to note that IBZs are meant to enhance the vegetated buffer zones and will cover less than 1% of the drained catchment. The implementation of IBZs has been proven to work in sandy to mixed-sandy loam soils with an organic carbon content higher than 2% of the soil dry mass, a slope greater than 3%, and along first and second order streams. The content of organic matter is important for the heterotrophic denitrification process that reduces nitrate to mainly N_2, however, it can be assumed that mostly fresh organic carbon from plant litter or root exudates is used by microbes (Zak et al., 2018).

The effectiveness of IBZs was determined to be 45 ± 12% for TN and 29 ± 60% for TP by Zak et al. (2018), and these percentages are similar to those of other drainage mitigation measures, such as different types of constructed wetlands (Hoffmann et al., 2020). To cope with surface runoff and sediment transport into streams, the IBZs must be placed at high-risk locations. IBZs are anticipated to be much more effective at trapping and ensuring longer-term retention of sediment and sediment-associated nutrients than traditional vegetated buffer strips installed along watercourses. To determine the optimal placement of buffer zones as sediment traps, the Water and Tillage Erosion Model (WaTEM) tool was applied in a study by Van Oost et al. (2000). This model enabled sediment redistribution in the landscape to be estimated at fine spatial resolution, including delivery to streams. Based on field runoff measurements, Zak et al. (2019) calculated that more than 292 kg of suspended sediment containing 1.39 kg P and 1.57 kg N entered an IBZ test site in western Denmark. Depending on the hydraulic load and vegetation growth at the test site, the removal rate for sediment and associated nutrients would eventually amount to 0.5–1.3 kg m^{-2}, 2–6 g P m^{-2} and 3–7 g N m^{-2}, respectively, during a 6-month winter period. The effectiveness of the IBZ as a sediment trap influences the maintenance effort required to cope with silt-up of the ditch pond of the IBZ. However, with an annual sediment growth rate in the pond of about 1 mm, several hundred years would need to pass before the aquatic part of the tested IBZ became completely filled with sediment.

An additional enhancement of riparian buffers is the implementation of saturated buffer zones (SBZs), which were first established and monitored in Iowa, US, in 2010 (Jaynes and Isenhart, 2019). The simple principle behind the SBZ is that drainage water from the field becomes reconnected to the non-cropped riparian zone. However, for an SBZ, drainage water and riparian buffer soil are reconnected by a buried, lateral perforated distribution pipe running parallel to the stream (0.5–1 m below the soil surface), which redirects the drainage water into the riparian zone and increases the groundwater table for a

Figure 6 Examples of four different buffer strip and buffer zone types. (a) Grass buffer strip; (b) buffer strip with trees; (c) integrated buffer zone and (d) saturated buffer zone.

few months during the drain season (Fig. 1). The effect on nitrate only was tested in the US, but the initial results from a pilot case study in Denmark show that removal efficiencies may be higher than 80% and 70% for TN and TP, respectively (Maagaard et al., 2022). These high removal rates are due to biogeochemical processes in the water-saturated soil, and it is therefore questionable to what extent the expected vegetation changes will alter the retention of sediment and associated nutrients from surface runoff from neighbouring agricultural fields. Figure 6 shows the four different types of buffer strips and buffer zones.

3 Experiences with buffer strips in Northwest Europe

3.1 Buffer strip regulation and implementation in Denmark

Buffer strips have been implemented in Danish law since 1992 when a 2-m mandatory buffer strip became required alongside all public streams, by order of the Watercourse Act. The main purpose of this 2 m buffer strip was to reduce direct fertiliser spillage in to streams during spreading and to prevent direct stream bank failures caused by the presence of large machinery near the water's edge. There is a general lack of evidence regarding the effectiveness of implementing buffer strips alongside stream field margins in terms of retaining sediment and sediment-bound nutrients resulting from soil erosion and surface runoff on arable fields, which has prompted a longer-term survey of soil erosion (rill erosion) and the effects of existing buffer strips along streams. The study encompassed 88–140 field slope units situated across Denmark during 1993-2000 (Kronvang et al., 2000). The results showed that average rill erosion on fields amounted to losses of 465 kg sediment per ha and 0.29 kg P per ha

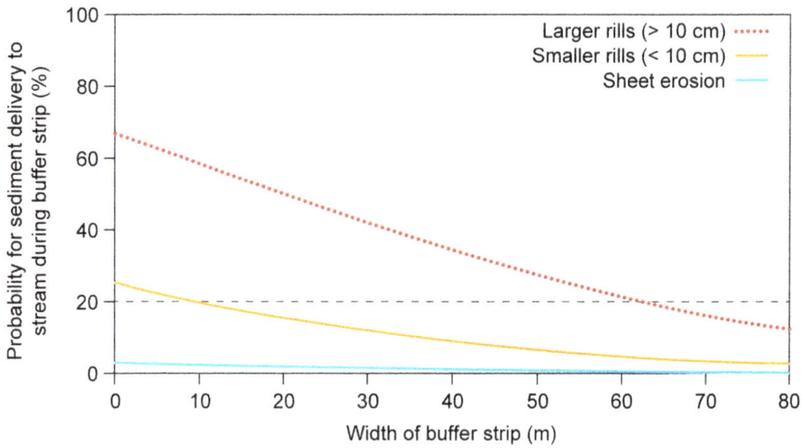

Figure 7 Probability plot for sediment delivery through the buffer strip depending on three erosion risk classes (two different rill sizes formed on neighbouring fields measured as rill diameter and sheet erosion) and the buffer strip width. Data comes from observations of 146 slope units during three winter periods in Denmark (1998/99 to 2000/01) (redrawn from Kronvang et al., 2005).

during a 6-year period (1993/94 to 1998/99). A survey of the buffer strips along the field slope units revealed that their effectiveness correlated highly to their width and the size of rills formed on the neighbouring fields (Laubel et al., 2000; Kronvang et al., 2005). This was also substantiated from observations documented in the international literature (Fig. 7).

In order to reduce sediment and nutrient losses from soil erosion and surface runoff, a proposal for voluntary implementation of 10 m buffer strips was introduced to farmers in 2003 as part of Denmark's 3rd Action Plan for the Aquatic Environment. The goal was to have 50 000 ha implemented before 2015, although a mid-way analysis conducted in 2009 revealed that only around 200 ha buffer strips had been implemented by farmers. The Green Growth Plan of 2011 mandated the introduction of 10 m buffer strips along all Danish streams and lakes larger than 100 m^2. The mandatory buffer strips were a cornerstone of the Danish government's efforts to implement the first phase of the WFD (2009–15), and the Buffer Zone Act represented the first attempt at large-scale implementation of the regulation based on a centrally coordinated assessment of effectiveness and environmental impact.

The Buffer Zone Act, which stipulated mandatory implementation of 10 m buffer strips by farmers, was adopted in June 2011 (Thorsøe et al., 2017). The Act should have been accompanied by a map detailing the buffer zone areas to farmers and was also meant to serve as a tool for disbursing compensation. However, the agricultural agency encountered difficulties in developing the maps as the input data were based on aerial photographs registered by the counties (until 2007) and municipalities. One of the challenges was that the Buffer Zone Act applies to all streams, not just the better charted classified streams. When the maps were finally published in Spring 2012, they were faulty and rejected by the agricultural industry. A new government was elected in 2011, after which the Buffer Zone Act was revised in June 2012, giving the public access to the buffer zones. Formerly in opposition, the Danish liberal party immediately resisted the Buffer Zone Act, which it perceived as a fundamental challenge to the farmers' private property rights (Thorsøe et al., 2017). The scientific evidence regarding the effectiveness of the 10 m wide buffer strips for nitrogen retention was strongly debated and questioned by a newly developed farmer union ('sustainable agriculture'), supported by international studies (Thorsøe et al., 2017).

In Spring 2014, a new agreement between the government and the opposition led to a simplification of the Buffer Zone Act, implementing a 50% reduction of the buffer zone area (Thorsøe et al., 2017). In June 2015, a new government took office, which withdrew the Buffer Zone Act in late 2015. Simultaneously, a new agricultural policy was adopted, which lifted some of the previous restrictions on fertiliser use and introduced a plan for large-scale implementation of new spatially differentiated regulation by 2019 (MFVM, 2015). Since 2016, farmers have been able to decide whether to implement buffer strips along streams instead of growing catch crops on their fields under the targeted catch crop regulation, with 1 ha buffer strips substituting 4 ha catch crops. The most recent regulation concerning buffer strips in Denmark will be introduced with the new EU Common Agricultural Policy (CAP) reform (2023–27), by which the establishment of 3 m buffer strips is mandatory where 2 m buffer strips already exist, in accordance with the Danish Watercourse Act of 1992.

3.2 Buffer strip regulation and implementation in Scotland

A growing understanding of catchment management throughout the 1990s in the UK and Europe culminated in the development of the EU Water Framework Directive at the turn of the century. These developments in knowledge and policy furthered our understanding of the inter-connected nature of land-based activities and the condition of waterbodies. In addition to addressing point source pollution from effluents, the term 'diffuse pollution' was expanded to encompass the ubiquitous background loss of soil and leakage of agrochemical and microbial contaminants from farmed land.

National water quality regulatory surveillance datasets collected near the mouth of Scotland's larger rivers (on a monthly basis, serving UK-wide purposes of land to sea fluxes) were less suited to storm-driven pollutants such as sediment and total phosphorus. In response, small-scale catchment studies using higher temporal resolution sampling (autosamplers and turbidity sensors) highlighted the fact that farmed and developed headwaters play a key role in distributing fine sediment, nitrogen, phosphorus and faecal contaminants, particularly when rains coincide with certain farm operations (Stutter et al., 2008 a,b).

These areas were deemed responsible for water quality deterioration in main stem river-receiving waters in Scotland, many of which had upstream oligotrophic sources and high conservation designations for protected species (including salmon, freshwater pearls, mussels and otters). During the early 2000s, the links between ecology and pollution, particularly fine sediment, in rivers became more pronounced and the potential for habitat loss and economic damage to fish stocks became increasingly apparent, pointing the focus towards addressing headwater farmland activities. However, whilst national regulations responded to the early implementation of the WFD, Catchment Partnerships developed workplans for management in a few select salmon fishing rivers, including grass and wooded buffer zones, implementing early morphological restoration and remediation of point sources, such as larger septic tanks.

On the often sloping, wet farmland of Scotland, the focus has been on implementing source controls to reduce fertiliser inputs, maintain appropriate livestock numbers and limit the exposure of bare, tilled soil during winter. The importance of interrupting pollution transport has been recognised, and vegetated buffer strips have become a central element in many developing catchment management plans without fully understanding their best practice usage and limitations. From around 2005, the national regulator, the Scottish Environment Protection Agency (SEPA), developed a leading approach with the farming and land-owning sector under an agency and land sector partnership group known as the Diffuse Pollution Management Advisory Group (DPMAG) (SEPA 2005, 2012). Over the course of a decade, DPMAG oversaw a strategy that included:

1. The development of a set of baseline principles of good practice for farming and forestry (termed General Binding Rules (GBRs));
2. Prioritisation of three phases of national catchments to examine and implement water quality improvements, consisting of agency staff walking stream corridors, mapping non-compliances with GBRs, and undertaking audits of farmyards; and
3. Offering advice and allowing time for farmers to uptake measures before the application of penalties.

This approach identified numerous non-compliances at stream margins, particularly concerning the cultivation and use of chemicals directly at the stream, storage of manure and siting of cattle feeders, and direct access of cattle to drinking water from the stream, leading to excess poaching and direct faecal contamination.

These observations cemented the case for a physical separation of the stream from the field, and vegetated, sometimes fenced, buffer zones of the minimum regulatory width (2 m from the bank top) were adopted by default. Breaches of the regulations were potentially enforceable and linked to penalties under the single farm payment (EU CAP, Pillar 1), although this was seldom applied. Meanwhile, the CAP Pillar 2 provided funding for agri-environmental initiatives, which necessitated separate applications in each case. The funding offered options for wider 6–10 m buffers for fencing and ongoing land-take payments, but uptake was negligible. It was notable that specific subcatchment targeting, led by catchment partnerships and local trusted farm advisors, was able to attract project money for species-led initiatives (e.g. the LIFE funded Pearls in Peril). These initiatives were effective in implementing wider buffers, riparian tree planting, extensive livestock exclusion fencing and alternative drinking arrangements. Such intensive regional examples of buffer zones became integrated with catchment research and allowed for a review of their effectiveness (Bergfur et al., 2012). However, the results varied, and whilst targeted stream reaches benefited from reduced sedimentation, slow readjustments to more natural channel forms, improved habitat and even public stream-side walking access, it was recognised that nitrate and phosphate water quality at the catchment scales showed limited, or at least delayed, results.

As the reporting requirements to the EU on national WFD implementation increased, it became evident that riparian grass, or vegetated buffer zones, were highly favoured in River Basin Management Plans. Achieving the required attainment of Good Ecological Status of waterbodies by the end of successive management planning cycles, however, was slower than anticipated. The physical condition of many streams and rivers, which had historically been deepened and straightened to facilitate field drainage, and continuing dredging to combat sedimentation, had hampered ecological improvements. As the first decade of WFD in Scotland came to a close, regulatory efforts focused on recognising and tackling morphological issues of streams and rivers, early stages of river restoration and improving flood management.

It took a reconsideration of management techniques in the riparian buffer space in the pursuit of multiple environmental and policy benefits to re-energise buffer zone discussions in Scotland, which happened in parallel with developments in England and across Ireland. This was necessary for two reasons. Firstly, whilst the initial 2 m vegetated margin provided physical separation between farming and development activities and streams through the no cultivation and no spread zones (under GBRs), farmers, advisors and regulators had little confidence that grass margins, even of moderate widths, would interrupt storm transfers of surface runoff, especially in clay-rich soils. Secondly, important new issues in terms of flood runoff and habitat were not sufficiently addressed by grass buffers alone.

Catchment scientists recognised that temporary water-holding measures of the Natural Flood Management (NFM) community shared similarities to the mechanics of retaining sediments and co-transported pollutants in bunds and sculpted ground (SEPA, 2015). They began to work together, developing concepts such as 'catchment sensitive engineering' to bring simple engineering principles into temporary water storage and settlement features (Hewett et al., 2020). The nature of low-lying field corners and less productive agricultural land meant that optimal feature-siting often occurred in wet riparian spaces where runoff converged. This brought about the development of accumulated flow tools based on digital elevation models (Stutter et al., 2020). Meanwhile, designs of buffers specific to wet ground (retaining runoff trapping ability but enhancing wetland processes for denitrification) were developed in partnership with Scandinavian researchers (Zak et al., 2019), and the development of wooded buffers was promoted (Stutter et al., 2019). An important step in the UK was an academic–regulatory partnership that developed the concept of '3D buffers', comprising riparian structural attributes from below ground (deep-rooted vegetation and good soil conditions for nutrient processing), to ground surface (sculpted ground profiles, vegetation roughness, wet-dry ground complexity), to the canopy (Stutter et al., 2020). Another large research initiative, SmarterBufferZ (funded by Irish EPA Research), has adopted the 'right measures, right place' principles for riparian buffers by increasing awareness of an expanded range of measure options for the riparian space and tools for guiding placement (Stutter et al., 2021b). Measure options include standard grass buffer space, wooded buffers, enhanced field surface and channel overbank runoff trapping, subsurface pathways including artificial drainage interception options and alteration of the channel space. These projects acknowledge the need for research approaches for identifying suitable riparian areas and raising awareness of the many different pollution trapping mechanisms available. The transition from planning to field walkovers and discussions between advisors and farmers is key to optimising buffers at the field to farm scale. Decision support tools play a huge role in this key implementation step.

There remains little coordinated action regarding the implementation of enhanced riparian buffer designs in Scotland. Some exceptions are specific catchment actions of tree planting for river shading, measures addressing pollution of higher level habitats, local NFM actions, and individual farmers self-organising to repair soil loss or nuisance runoff. Wider grass margins are rarely implemented, and narrow grass buffers are failing to achieve optimum water quality, habitat and wider objectives. It is hoped that the collaborations that have taken place across the British Isles and Scandinavia over the past 10 years will stimulate a greater range of surface and subsurface flow interception designs to gain favour in specific catchment management actions. It is further anticipated that some of these, such as wooded buffers, will be adopted in new agri-environment schemes that are developing post-Brexit. A position statement detailing relevant work towards this goal has been developed (Stutter and Wilkinson, 2022) to strengthen awareness of the increasing scope of designs, data and tools to promote their effective placement. It should be emphasised that while the potential for tailored designs of riparian buffer zones to suit localised field and margin conditions has been extensively researched, implementation remains low. Demonstrations and trials are needed to improve acceptability and practical knowledge to guide their deployment.

3.3 Buffer strip regulation and implementation in Norway

The main impetus behind the first study on reducing pollution from soil erosion in agricultural areas in Norway was the eutrophication of the country's largest lake, Lake Mjøsa. The severe state of the lake in the 1970s led to an extensive period of environmental planning and action during 1973–82, in which several measures were introduced to reduce nutrient losses. At the beginning, the primary nutrient sources, including sewage, phosphorus-rich detergents, industrial effluents and the application of fertiliser and manure to fields, were targeted (Ministry of Environment, 1979), whilst runoff from cultivated land was considered challenging to address (Brobakk and Melås, 2020). At the end of the 1980s, an official from the County Governor's office in Oslo and Akershus announced plans to reduce nutrient losses to Lake Mjøsa. One of the proposed measures was the implementation of vegetated buffer zones (Bogen et al., 1989). Meanwhile, in a report to the Nordic Council of Ministers (Svendsen and Kronvang, 1991), the only reference to work on vegetated buffer zones was a literature study that focused on experiences in other countries (Bechmann, 1990). Nonetheless, interest in buffer zones began to grow, which can also be attributed to the requirements of the North Sea Agreement, signed in 1987, in which Norway agreed to reduce phosphorus and nitrogen losses to the North Sea by 50% compared to 1985 levels. Pressure mounted on various sectors to lower nutrient levels, leading to closer scrutiny of the primary sources of phosphorus input.

The agricultural sector was assigned the important task of identifying the proportion of phosphorus deriving from natural bank erosion of phosphorus-rich marine clay soils and the proportion deriving from agricultural activities. Projects to address this issue were commissioned and financed by the Agricultural Office of the County Municipality of Akershus and Oslo in the late 1980s/early 1990s. The chosen method was to estimate sediment sources in a designated case study river. The study concluded that losses from natural bank and ravine erosion amounted to around 55%, whilst agricultural activities caused about 45% of total sediment transport (Bogen et al., 1989, 1993). It should be noted that the case study river (River Leira in East Norway) runs through steep ravine valleys, and the study is therefore not representative of all clay-soil areas in Norway. The first experimental plot study on the retention capacity of vegetated buffer zones was initiated and later reported by Syversen (2002). These studies documented that retention in 5–10 m wide buffer zones was 55–97% for particles, 42–96% for total phosphorus and 27–81% for total nitrogen, but this was of course true only of the slope and soil types found at this particular plot site. These were the only experimental results on the effectiveness of buffer zones in Norway until renewed interest emerged about a decade later (Skarbøvik and Blankenberg, 2013; Blankenberg and Skarbøvik, 2020).

Vegetated buffer strips along streams in agricultural areas are controlled by various Norwegian regulations and economic incentives, some of which can, to a certain degree, be contradictory. Norway transposed the EU WDF in December 2006. Of the laws that constitute the legal basis for this regulation, the Water Resources Act (2001) is of particular interest in relation to buffer strips since Section 11 of the Act states that 'a limited, natural vegetation zone to reduce runoff and be a habitat for plants and animals, should be maintained along water courses with annual water flow'. The phrase 'maintain' rather than 'establish' is interesting since the much older Land Act (1995, Section 9) says that 'Cultivated land must not be used for other purposes than agricultural production'. However, it is not entirely prohibited to change the land use in cultivated soil; the second part of this paragraph states that the Agricultural Ministry can, in special circumstances, give permission for uses of agricultural land other than food production following a comprehensive evaluation. The background for this paragraph is a need to protect soils suitable for agricultural production in Norway – only 3.5%, or about 1 million ha of the total land area, is currently under cultivation, of which about 30% is suitable for cereal production.

An important regulation under the Land Act (1995) is the Regulation of Production Grants (2014). This instructs that production grants can be allocated per area of cultivated land but with various requirements, one of which is that the farmers must maintain a natural buffer zone between agricultural land and streams with annual water flow. The zone should be wide enough to prevent runoff to water bodies under normal water flow and should be at least 2 m wide. The extent of the 2 m wide buffer strip is measured horizontally from the streams' normal water level, and for large rivers, it can be argued that these 2 m are hardly more than the stretch from the river to the top of the riverbank. Since the minimum width is 2 m, many farmers do not extend this zone any further. This highlights an important shortcoming of the Norwegian regulations, i.e. the legal requirements governing the width of natural buffer zones in agricultural areas. However, when trees and other natural vegetation are already growing along the banks, it is not permitted to remove them within a width of 6 m, as stated in the Regulation of Cultivation of New Land (2021).

In Norway, agricultural regional environment plans are the main economical instrument for implementing environmental mitigation measures in agriculture. The first time that vegetated buffer zones were eligible for economic incentives in Norway was in 2005 (Gundersen et al., 2009). Whilst each region can have different economic incentives, the overarching national environment plan (Directorate of Agriculture, 2021) dictates that to receive subsidies for buffer zones with grass production bordering agricultural fields, the total width of the zone must be 8 m, including the 2 m regulated by the production grants described earlier. In meadows, a similar economic incentive applies for unfertilised buffer zones. For both varieties, no fertilisation or pesticides must be applied in the zones, and the grass should be harvested or grazed to prevent leakage of nutrients when it withers during autumn and winter.

As can be seen from the earlier discussion, there is no other countrywide law that stipulates the width of buffer zones than the 2 m required to receive production grants, and even this can be deemed voluntary, albeit with a strong economic incentive. However, in a few catchments, local regulations requiring implementation of buffer zones were introduced in 2009 and were fairly successful as many farmers did indeed introduce buffer zones (Skaalsveen et al., 2021; Bechmann et al., 2022). However, the regulation was rescinded in 2013 following a decision by the agricultural authorities. Today, the eutrophication and poor state of the Oslo Fjord have led to growing awareness and, therefore, new regional regulations in parts of the county of Oslo & Viken, ratified in 2023. This stipulates that all agricultural areas within these parts of the county shall have a buffer zone of either 6 m grass production or 20 m unploughed field (stubble) in addition to the 2 m required to receive production grants. More counties are currently preparing similar regulations, which might indicate a move from mere economic incentives to more nationwide legal regulations of buffer zones. This, however, remains to be seen.

4 Conclusion and future trends

The effectiveness of buffer strips along streams and lakes for the purpose of retaining sediment and sediment-associated nutrients has been documented in numerous studies throughout the world. The most recent evidence highlighted in this chapter shows that the trapping efficiency of buffer strips is mainly linked to the width of buffer strips and the clay content of buffer strip soils, with the slope of the buffer strip of minimal importance. The role of vegetation in the buffer strips requires further investigation.

Doubts remain regarding the effectiveness of buffer strips for removal of dissolved nutrients such as nitrate and dissolved phosphorus forms. Accordingly, further studies elucidating the role of buffer strips and distinguishing the different flow paths through the buffer strip are required to enhance our knowledge and to help achieve best options/tools for dissolved nutrient management.

Optimal design guidelines and demonstration of riparian buffer zones localised at different field margin conditions are also needed, including further testing of the capture of tile drainage water from agricultural fields by the newly developed IBZs and SBZs.

Studies aimed at investigating the multifunctionality of buffer strips and buffer zones in order to achieve future goals for the environment, climate and biodiversity are also in demand, with particular focus on the effects of wooded buffer strips.

The definition of buffer strips remains uncertain and varies among countries. As described in this chapter, in Norway buffer strips start horizontally from the normal water level within the stream channel, whilst in Scotland and Denmark, they start from the edge of the stream banks. This difference in definition may seem insignificant, but is actually not the case as the narrow buffer strips often used in legislation (2–4 m) will have minimal effective impact if part of the buffer strip consists of the stream bank.

New mapping models and tools capable of assisting catchment managers in targeted placement of buffer strips and buffer zones are widely accepted and used, such as the WATEM model for soil erosion and surface runoff processes used in Denmark. The predicted impact of climate change on extreme rain events will likely increase the importance of soil erosion and surface runoff for transport of sediment and nutrients to surface waters. Model tools that can assist in the targeted placement of buffer strips in catchments therefore remain necessary.

The costs of buffer strips relative to the need for food production do of course have to be taken into account at regional and national level, and farmers, citizens and politicians must analyse this prior to implementing buffer strip regulations. In regions and countries with low proportions of agricultural land, stakeholders and policy makers may be reluctant to adopt buffer strips simply due to the limited land area available for agricultural production. The costs and benefits of placing targeted buffer strips have to be calculated and compared with the costs and benefits of continuing agricultural production, including the multifunctional benefits of buffer strips. This will require further development of economic models, including the analysis of wider biodiversity gains.

Lastly, the attitude of farmers to buffer strips and buffer zones as a mitigation measure is vital, as they are the principal stakeholders that have to be convinced in order to implement them on their farms. Solid scientific documentation of the effects of buffer strips, together with development of demonstration material (fact sheets, videos etc.) and regional demonstration sites are required if farmers are to be convinced of the importance of buffer strips for the environment, climate and biodiversity.

5 Acknowledgements

This work was supported by the Nordic Centre of Excellence BIOWATER, funded by NordForsk under Project No. 82263, the EU HORIZON project NORDBALT-ECOSAFE Grant Agreement No. 101060020, the Research Council of Norway (Contract No. 342631/L10) and the Scottish Government Rural and Environment Science and Analytical Services (RESAS).

6 Where to look for further information

A newly finished research project under the Environment Agency, UK '3D buffer strips designed to deliver more for the environment' (https://www.gov.uk/government/publications/3d-buffer-strips-designed-to-deliver-more-for-the-environment) has considered how we can improve the effectiveness of riparian buffer zones to help tackle agricultural pollution.

7 References

Abu-Zreig, M., Rudra, R. P., Whiteley, H. R., Lalonde, M. N. and Kaushik, N. K. 2003. Phosphorus removal in vegetated filter strips. *Journal of Environmental Quality* 32(2): 613–619.

Bechmann, M. 1990. Vegetasjonssoner langs vassdrag. Virkning på forurensning. JORDFORSK-rapport. *AS-NLH* 36.

Bechmann, M., Skarbøvik, E., Turtumøygard, S. and Isdahl, C. R. 2022. Trender i tiltaksgjennomføring og påvirkning på vannkvalitet i Morsa. (Trends in implementation of environmental measures, and the effect on water quality in the Morsa river Basin). *NIBIO Report* 155 (8): 37. https://doi.org/10.1038/s41598-020-79473-x.

Bergfur, J., Demars, B. O. L., Stutter, M. I., Langan, S. J. and Friberg, N. 2012. The Tarland Catchment Initiative and its effect on stream water quality and macroinvertebrate indices. *Journal of Environmental Quality* 41(2): 314–321.

Blankenberg, A. B. and Skarbøvik, E. 2020. Phosphorus retention, erosion protection and farmers' perceptions of riparian buffer zones with grass and natural vegetation: case studies from South-Eastern Norway. *Ambio* 49(11): 1838–1849. https://doi.org/10.1007/s13280-020-01361-5.

Bogen, J., Moen, E. and Sandersen, J. 1989. Forurensing som følge av leirerosjon og betydning av erosjonsforebyggende tiltak. Rapport Forbygningsavdelingen NVE (In Norwegian), 54 pp. (Pollution deriving from erosion in clay soils and the effect of erosion reducing measures).

Bogen, J., Berg, H. and Sandersen, F. 1993. Forurensning som følge av leirerosjon og betydningen av erosjonsforebyggende tiltak: Sluttrapport. *NVE Rapport* 21; 96 pp. (Pollution deriving from erosion in clay soils and the effect of erosion reducing measures; Final report; In Norwegian).

Borin, M., Vianello, M., Morari, F. and Zanin, G. 2005. Effectiveness of buffer strips in removing pollutants in runoff from a cultivated field in North-East Italy. *Agriculture, Ecosystems and Environment* 105(1–2): 101–114.

Brobakk, J. and Melås, A. M. 2020. Dagens klimautfordringer og tidligere miljøtilpasninger i Norsk jordbruk – hva kan vi lære? 7/2020 Ruralis - Institutt for Rural- og regionalforskning, 47 pp. *Rapport*. (Today's climate challenges and earlier environmental adaptations in Norwegian agriculture – what can be learned?; in Norwegian).

Cole, L. J., Stockan, J. and Hellliwell, R. 2020. Managing riparian buffer strips to optimize ecosystem services: a review. *Agriculture, Ecosystems and Environment* 296: 1–12.

Daniels, R. B. and Gilliam, J. W. 1996. Sediment and chemical load reduction by grass and riparian filters. *Soil Science Society of America Journal* 60(1): 246–251.

Dillaha, T. A., Sherrard, J. H., Lee, D., Mostaghimi, S. and Shanholtz, V. O. 1988. Evaluation of vegetative filter strips as a best management practice for feed lots. *Journal of Water Pollution Control Federation* 60: 1231–1238.

Directorate of Agriculture 2021. Regionalt miljøtilskudd i jordbruket (RMP) – kommentarer til regelverk. *Rundskrivnummer* 2021/13. Available at: https://www.landbruksdirektoratet.no/nb/jordbruk/ordninger-for-jordbruk/regionalt-miljotilskudd-rmb/regionalt-miljotilskudd-rmp--kommentarer-til-regelverk/kapittel-6.avrenning-til-vann.

Dorioz, J. M., Wang, D., Poulenard, J. and Trévisan, D. 2006. The effect of grass buffer strips on phosphorous dynamics – a critical review and synthesis as a basis for application in agricultural landscapes in France. *Agriculture, Ecosystems and Environment* 117(1): 4–21.

Gericke, A., Nguyen, H. H., Fischer, P., Kail, J. and Venohr, M. 2020. Deriving a bayesian network to assess the retention efficacy of riparian buffer zones. *Water* 12(3): 617; https://doi.org/10.3390/w12030617.

Gundersen, G. I., Snellingen Bye, A. S., Berge, G., Hoem, B. and Knudtsen, S. S. 2009. Jordbruk og miljø. Tilstand og utvikling. *Statistics Norway* (Statistisk Sentralbyrå, Society of Systematic Biologists) Rapportar 2009/37. 102 pp. (Agriculture and Environment. State and Development; in Norwegian). Available at: https://www.ssb.no/a/publikasjoner/pdf/rapp_200937/rapp_200937.pdf.

Haddaway, N. R., Brown, C., Eales, J., Eggers, S., Josefsson, J., Kronvang, B., Randall, N. P. and Uusi-Kämppä, J. 2018. The multifunctional roles of vegetated strips around and within agricultural fields. *Environmental Evidence* 7: 14.

Heinen, M., Noij, I. G. A. M., Heesmans, H. I. M., van Groenigen, J. W., Groenendijk, P. and Thissen, J. T. N. M. 2012. A novel method to determine buffer strip effectiveness on deep soils. *Journal of Environmental Quality* 41(2): 334–347.

Hewett, C. J. M., Wilkinson, M. E., Jonczyk, J. and Quinn, P. F. 2020. Catchment systems engineering: an holistic approach to catchment management. *Wires Water* 7(3): e1417. Available at: https://wires.onlinelibrary.wiley.com/doi/abs/10.1002/wat2.1417.

Hille, S., Graeber, D., Kronvang, B., Rubaek, G. H., Onnen, N., Molina-Navarro, E., Baattrup-Pedersen, A., Heckrath, G. J. and Stutter, M. I. 2019. Management options to reduce phosphorus leaching from vegetated buffer strips. *Journal of Environmental Quality* 48(2): 322–329. https://doi.org/10.2134/jeq2018.01.0042.

Hoffmann, C. C., Kjaergaard, C., Uusi-Kamppa, J., Hansen, H. C. and Kronvang, B. 2009. Phosphorus retention in riparian buffers: review of their efficiency. *Journal of Environmental Quality* 38(5): 1942–1955.

Hoffmann, C. C., Zak, D., Kronvang, B., Kjaergaard, C., Carstensen, M. V. and Audet, J. 2020. An overview of nutrient transport mitigation measures for improvement of water quality in Denmark. *Ecological Engineering* 155. https://doi.org/10.1016/j.ecoleng.2020.105863.

Jaynes, D. B. and Isenhart, T. M. 2019. Performance of saturated riparian buffers in Iowa, USA. *Journal of Environmental Quality* 48(2): 289–296. https://doi.org/10.2134/jeq2018.03.0115.

Kronvang, B., Laubel, A., Larsen, S. E., Iversen, H. L. and Hansen, B. 2000. The role of erosion and sediment transport in nutrient and contaminant transfer. In: Stone, M. (Ed.) Proceedings of the Symposium Held at Waterloo, Canada, July 2000). *IAHS Publication* 263: 67–73.

Kronvang, B., Laubel, A., Larsen, S. E., Andersen, H. E. and Djurhuus, J. 2005. Buffer zones as a sink for sediment and phosphorus between the field and stream: Danish field experiences. *Water Science and Technology* 51(3-4): 55–62.

Kronvang, B., Audet, J., Baattrup-Pedersen, A., Jensen, H. S. and Larsen, S. E. 2012. Phosphorus load to surface water from bank erosion in a Danish lowland river basin. *Journal of Environmental Quality* 41(2): 304–313. https://doi.org/10.2134/jeq2010.0434.

Land Act 1995. *Lov om jord (Jordlova).* (in Norwegian). Available at: https://lovdata.no/dokument/NL/lov/1995-05-12-23.

Laubel, A., Kronvang, B., Svendsen, L. M., Pedersen, M. L. and Larsen, S. E. 2000. Danish buffer strips as sediment and phosphorus traps. *Verhandlungen der Internationale Vereinigung für Theoretische und Angewandte Limnologie* 27(3): 1536–1540.

Lee, K. H., Isenhart, T. M., Schultz, R. C. and Mickelson, S. K. 2000. Multi-species riparian buffers trap sediment and nutrients during rainfall simulations. *Journal of Environmental Quality* 29(4): 1200–1205.

Liu, X., Zhang, X. and Zhang, M. (2009). Major factors influencing the efficacy of vegetated buffers on sediment trapping: a review and analysis. *Journal of Environmental Quality* 37(5): 1667–1674.

Maagaard, A., Petersen, R. J., Kronvang, B., Ovesen, N. B., Audet, J., Hoffmann, C. C. and Zak, D. H. 2022. Innovations platform for virkemidler til begrænsning af drænvand.

Foreløbige resultater vedrørende fjernelse af kvælstof og fosfor i en mættet randzone. Aarhus Universitet, DCE, Dansk Center for Miljø og Energi, 23 s. – Fagligt notat nr. 2022|25.

MFVM 2015. *Aftale om Fødevare- og landbrugspakke*. Copenhagen. Ministry of Environment and Food, Denmark.

Ministry of Environment 1979. The Mjøsa campaign. Status Report. The Norwegian Ministry of Environment, 100 pp. Available at: https://web.archive.org/web/20160309201237/http://fylker.miljostatus.no/Global/Hedmark/vannforurensning/Statusrapport_Mjosaksjonen_46sDS-file2668.pdf.

Moset, V., Hille, S., Rubæk, G. H., Møller, H. B., Wahid, R. and Baattrup-Pedersen, A. 2019. Indicators of biomass and methane yields in vegetated buffer strips. *Journal of Cleaner Production* 210: 907–915.

Noij, I. G. A. M., Heinen, M., Heesman, H. I. M., Thissen, J. T. N. M. and Groenendijk, P. 2012. Effectiveness of unfertilised buffer strips for reducing nitrogen loads from agricultural lowland to surface waters. *Journal of Environmental Quality* 41(2): 322–333.

Regulation of Cultivation of New Land (2021). *Forskrift om nydyrking* (in Norwegian). Lovdata. Available at: https://lovdata.no/dokument/SF/forskrift/1997-05-02-423.

Regulation of Production Grants 2014, 2021. *Forskrift om produksjonstilskudd og avløsertilskudd I jordbruket* (in Norwegian). Lovdata. Available at: https://lovdata.no/dokument/SF/forskrift/2014-12-19-1817?q=produksjonstilskudd.

SEPA 2005. *River Basin Planning Strategy for the Scotland River Basin District*. Scottish Environment Protection Agency. Available at: www.sepa.org.uk/media/37453/rbmp_planning-strategy_scotland.pdf. Accessed on February 2023.

SEPA 2012. Rural diffuse pollution plan for Scotland. Available at: www.sepa.org.uk/media/37557/rural-diffuse-pollution-plan-scotland.pdf. Accessed on February 2023.

SEPA 2015. *NFM Handbook*. Available at: https://www.sepa.org.uk/ media/163560/sepa-natural-flood-management handbook1.pdf.

Skaalsveen, K., Øverland, J. I., Bechmann, M., Maurset, M. U., Kvakkestad, V., Wiik, J., Evju, I. and Holm, H. N. 2021. Barrierer og muligheter for gjennomføring av vannmiljøtiltak (in Norwegian). NIBIO Report 8/38/2022; 52 pp. (Barriers and possibilities for implementing environmental measures for water quality).

Skarbøvik, E. and Blankenberg, A.-G. B. 2013. Vurdering av kanterosjon og vegetasjonssoner i Haldenvassdraget oppstrøms Bjørkelangen (in Norwegian). *Bioforsk rapp* 8(161), 53 pp. (Evaluation of bank erosion and vegetated buffer zones in the Haldenvassdraget Catchment upstream of Lake Bjørkelangen).

Søvik, A. K., Syversen, N., Blankenberg, A. G. B. and Mæhlum, T. 2012. Retention of agricultural surface runoff in a cold climate vegetative buffer zone-effect of vegetation and season. *Journal of Water Management and Research* 68: 86–96.

Stutter, M. I., Langan, S. J. and Cooper, R. J. 2008a. Spatial and temporal dynamics of stream water particulate and dissolved N, P and C forms along a catchment transect, NE Scotland. *Journal of Hydrology* 350(3–4): 187–202.

Stutter, M. I., Langan, S. J. and Cooper, R. J. 2008b. Spatial contributions of diffuse inputs and within-channel processes to the form of stream water phosphorus over storm cycles. *Journal of Hydrology* 350(3–4): 203–214.

Stutter, M. I., Langan, S. J. and Lumsdon, D. G. 2009. Vegetated buffer strips can lead to increased release of phosphorus to waters: a biogeochemical assessment of the mechanisms. *Environmental Science and Technology* 43(6): 1858–1863.

Stutter, M., Chardon, W. and Kronvang, B. 2012. Riparian buffer strips as a multifunctional management tool in agricultural landscapes: introduction. *Journal of Environmental Quality* 41(2): 297–303.

Stutter, M., Kronvang, B., Ó hUallacháin, D. and Rozemeijer, J. 2019. Current insights into the effectiveness of riparian management, attainment of multiple benefits, and potential technical enhancement. *Journal of Environmental Quality* 48(2): 236–247.

Stutter, M., Wilkinson, M. and Nisbet, T. 2020. 3D buffer strips: designed to deliver more for the environment. Bristol, UK. Available at: https://www.gov.uk/government/publications/3d-buffer-strips-designed-to-deliver-more-for-the-environment.

Stutter, M., Costa, F. B. and Ó hUallacháin, D. 2021a. The interactions of site specific factors on riparian buffer effectiveness across multiple pollutants: a review. *Science of the Total Environment* 798. https://doi.org/10.1016/j.scitotenv.2021.149238.

Stutter, M., Baggaley, N., Ó hUallacháin, D. and Wang, C. 2021b. The utility of spatial data to delineate river riparian functions and management zones: a review. *Science of the Total Environment* 757: 143982.

Stutter, M. and Wilkinson, M. 2022. Can improved design concepts for riparian buffer measures and placement improve uptake and best practice in Scotland? Scotland's Centre of Expertise for Waters (CREW). Project CRW2020_03. Available at: https://www.crew.ac.uk/sites/www.crew.ac.uk/files/publication/CRW2020_03_Buffer_Design_Placement_FINAL.pdf. Accessed on February 2023.

Svendsen, L. M. and Kronvang, B. 1991. Phosphorus in the Nordic Countries - methods, bioavailability, effects and measures. *NORD* 47: 201.

Syversen, N. 2002. Cold climate vegetative buffer zones as filters for surface agricultural runoff. Retention of soil particles, phosphorus and nitrogen. Dr. scient. thesis 2002:12. Agricultural University of Norway.

Thorsøe, M. H., Graversgaard, M. and Noe, E. 2017. The challenge of legitimizing spatially differentiated regulation: experiences from the implementation of the Danish buffer zone act. *Land Use Policy* 62: 202–212. https://doi.org/10.1016/j.landusepol.2016.12.030.

Uusi-Kämppä, J., Turtola, E., Närvänen, A., Jauhiainen, L. and Uusitalo, R. 2012. Phosphorus mitigation during spring time runoff by amendments applied to grassed soil. *Journal of Environmental Quality* 41(2): 420–426.

Van Oost, K., Govers, G. and Desmet, P. 2000. Evaluating the effects of changes in landscape structure on soil erosion by water and tillage. *Landscape Ecology* 15(6): 577–589. https://doi.org/10.1023/A:1008198215674.

Water Resources Act 2001. *Lov om vassdrag og grunnvann (vannressursloven)* (in Norwegian). Lovdata. Available at: https://lovdata.no/dokument/NL/lov/2000-11-24-82.

Zak, D., Kronvang, B., Carstensen, M. V., Hoffmann, C. C., Kjeldgaard, A., Larsen, S. E., Audet, J., Egemose, S., Jorgensen, C. A., Feuerbach, P., Gertz, F. and Jensen, H. S. 2018. Nitrogen and phosphorus removal from agricultural runoff in integrated buffer zones. *Environmental Science and Technology* 52(11): 6508–6517. https://doi.org/10.1021/acs.est.8b01036.

Zak, D., Stutter, M., Jensen, H. S., Egemose, S., Carstensen, M. V., Audet, J., Strand, J. A., Feuerbach, P., Hoffmann, C. C., Christen, B., Hille, S., Knudsen, M., Stockan, J., Watson, H., Heckrath, G. and Kronvang, B. 2019. An assessment of the multifunctionality of integrated buffer zones in Northwestern Europe. *Journal of Environmental Quality* 48(2): 362-375.

Chapter 8

The effects of no-till and related practices in preventing soil erosion

Michael Kucera, Formerly USDA Natural Resources Conservation Service (USDA-NRCS)/ National Soil Survey Center, USA

1 Introduction

This chapter will describe the use of no-till (NT) and related practices, and the importance of a fully integrated cropping and residue management system to maintain soil health, productivity, optimize soil functions such as nutrient and water cycling and, in particular, to reduce erosion. Use of techniques such as NT allow maintenance of continuous crop residue and/or canopy cover of soil, which are arguably the most important and practical measures to reduce both water and wind erosion. This is especially important to make farms and soils more resilient to more extreme drought and heavy rainfall events caused by climate change. There is now a wealth of research both on the mechanics of soil erosion and linking NT practices to improving the physical as well as chemical and biological health of soils, especially when combined with other practices such as crop residue retention (Rickson, 2018; Hamm and Manter, 2022; Gollany, 2022).

The chapter starts with definitions of key terms and practices. These are based in part on definitions in the United States Department of Agriculture-Natural Resources Conservation Service (USDA-NRCS) National Handbook

http://dx.doi.org/10.19103/AS.2023.0131.15

of Conservation Practices (https://www.nrcs.usda.gov/resources/guides-and-instructions/conservation-practice-standards).

Conventional tillage (CT) (abbreviated to Con T in this chapter to differentiate it from Conservation tillage (CT) which is discussed below) is an all-encompassing term that includes all tillage and planting systems that leave less than 30% of the soil surface covered with crop residue after planting. It includes any full width tillage system or inversion tillage that disturbs the entire surface of the soil. It can include a single pass with a moldboard plow (inversion of the soil) and may or may not be followed by a secondary tillage operation such as disking and/or harrowing. Conventional tillage can include mulch tillage systems that consist of multiple tillage passes after harvest that leave little protective ground cover after planting the next crop. It can also include tillage events prior to seeding a forage or pasture crop, where the soil is firmed with a roller or cultipacker.

NT is also known as zero-till, direct seeding or slot planting. NT is defined in the USDA-NRCS National Handbook of Conservation Practices as 'Limiting soil disturbance to manage the amount, orientation and distribution of crop and plant residue on the soil surface year around'. To compare different practices, the USDA-NCRS has developed Soil Tillage Intensity Rating (STIR) based on factors such as tillage type, depth of tillage operation and amount of soil left undisturbed and using the Revised Universal Soil Loss Equation Version 2 (RUSLE 2) to measure erosion impact. NT operations require a score of 30 or less (preferably 15). NT applies to cropland and allows for minimal in-row operations to disturb soil during planting and fertilizer application (Fig. 1). NT planting operations might include direct drills for planting seed, row-cleaners, coulters, opening disks and closing wheels for soil management. Fertilizer placement may occur at planting or using injection or surface application. In NT systems, the soil is undisturbed by tillage during the entire year following harvest of the previous crop. Crop residues left on the soil surface may be disturbed in strips up to one fourth of the row width for planting or drilling seed and fertilizer injection. It is also important to assess whether row cultivation will be used with a specific type of tillage system or not since it will reduce levels of residue left on the surface. It is also important to take into account whether residue is removed after harvest through burning, baling, grazing, or other residue removal methods which reduce ground cover. The level of soil disturbance is greater with NT systems with row cultivation versus those without.

Conservation tillage (CT) encompasses a wide range of reduced tillage and planting systems that leave a protective residue on the soil surface. CT has been defined by the Conservation Tillage Information Center (CTIC) in terms of its outcomes: that plant residue cover of the soil surface should be at least 30% (CTIC, 2017). Reduced tillage has been defined as systems that operate at

Figure 1 No-till corn planting into wheat stubble in Nebraska. Source: Photo courtesy of Michael Kucera.

shallower depths or at lower intensity than conventional ploughing, with some defining shallower depths at 10–20 cm (4–8 inches) compared, for example, to deep inversion tillage using moldboard ploughs at depths of 20–40 cm (8–14 inches) (Krauss et al., 2022). CT systems include techniques such as NT, strip-till, mulch tillage, ridge tillage and other systems that leave at least 30% or more cover (https://www.ctic.org/) .

Strip-till is also referred to by other names such as zone or row tillage (Fig. 2). It is often used interchangeably with NT but is distinctly different. Strip-till retains a number of the benefits of NT but disturbs a portion of the row or zone using tillage practices only where the next crop will be planted. The space between the rows is left undisturbed and covered with residue. Strip-till can be undertaken during planting or with pre-plant zone tillage to incorporate fertilizer to improve the seedbed environment, disturbing the soil only in a narrow zone typically 15–20 cm (6–8 inches) wide and/or less than 50% or the row area. Tillage depth can be 2.5–5 cm (1–2 inches) for fertilizer incorporation or up to 15–20 cm (6–8 inches deep). USDA-NRCS places NT and strip-till into the same conservation practice standard but with a higher allowable STIR of 60 for CT techniques such as strip-till.

In **mulch tillage (MT),** the entire soil surface is disturbed by non-inversion tillage (full-width tillage) whereby crop residues are mixed with the surface soil (Fig. 3). A wide variety of tillage implements can be used such as disks, chisels, field cultivators, vertical tillage or other non-inversion tillage methods. Depending on the extent of non-inversion tillage residue levels may be less than or greater than 30%. Mulch tillage systems that leave more than 30% cover

Figure 2 Strip Till fertilizer application into corn stubble photo courtesy of AgFax.

Figure 3 Mulch tillage of corn stubble in Nebraska. Source: Photo courtesy of Michael Kucera.

after planting are considered CT systems and those that leave less than 30% are considered conventional tillage systems.

In **ridge-till,** specialized planters and cultivators are used to form and retain permanent ridges on which cash crops are grown. Crops are planted on the top of the ridge after removing residue, which is left between ridges. Row cultivation is used to form and maintain ridges and to manage weeds.

The role of NT is partly to improve soil physical properties, including reducing sheet, rill and wind erosion, maintaining or increasing soil quality and soil organic matter (SOM) content, and increasing water use and precipitation storage efficiency. It is important to note that NT or conservation tillage by itself will not always reduce erosion to desired levels. NT must be coupled with other complementary practices such as crop residue retention, use of high residue crops, diverse crop rotations, cover crops and other management measures that lead to high levels of residue cover left on the soil surface to reduce raindrop impact, soil detachment and splash movement of soil particles. For example, NT or CT systems that utilize low residue crops such as silage crops, soybeans, vegetables, edible beans and canola or when excessive stover is removed by grazing, baling or other removal on a regular basis will not provide adequate erosion control. In addition, other erosion control practices such as strip cropping, terraces, contour farming or grassed waterways may also be important especially on steeper slopes.

Some of the key benefits for erosion control by maximizing crop residue on the soil surface are as follows:

- Crop residue left on the soil surface especially standing residue, reduces wind energy at the soil surface reducing wind erosion forces and soil displacement from creep and saltation of soil particles.
- Crop residue left on the soil surface absorbs the energy from raindrop impact reducing soil particle detachment and splash.
- By protecting the soil surface, surface crusting is also reduced, improving infiltration and decreasing runoff which transports sediment.
- This protective effect conserves soil and water and reduces risks to the environment from runoff.
- Residue cover conserves water by reducing evaporation from the soil surface.
- Decaying residue improves organic matter in soils feeds soil microbes and earthworms, promotes cycling of nutrients, improves soil aggregation and builds soil structure (which reduces erosion risk).

Other benefits include reducing greenhouse gas (GHG) emissions from soil (through minimal soil disturbance) and reduced fuel and labor costs (due to not undertaking conventional ploughing). Crop residue must be properly managed year-round to provide erosion control and other benefits without interfering with crop production. A well planned system accounts for impacts on nutrient, pest and water management.

2 The effects of no-till practices on soil properties related to erosion risk

Degradation of agricultural soils caused by excessive tillage has spurred interest in NT cropping systems. NT cropping creates the physical conditions of a relatively undisturbed soil that leaves residue on the surface, potentially making the soil less susceptible to wind and water erosion (Baker et al., 2007). In recent decades, investigations documenting spatially averaged comparisons between tillage practices, usually contrasting 'snap shots' of various tillage systems versus NT, have become relatively common. NT treatments have been compared with various tillage practices (e.g. conventional tillage, NT, mulch-till and ridge-till) under a range of conditions, soils and cropping systems with mixed results on changes to soil physical properties (such as porosity, bulk density (BD), penetration resistance, aggregate stability and saturated hydraulic conductivity). Table 1 provides a summary of a range of key studies on short- versus long-term NT, taking account of factors such as different soil temperature and moisture regimes, soil types and other related agronomic practices that impact NT effectiveness. It is important to note that many studies shown in Table 1 are short term (less than 5 years) or medium term (less than 12 years) which may not fully reflect gradual soil physical property improvements over time.

The impact of NT on soil physical properties varies in many studies and is somewhat dependent on climate, amount of disturbance, crop rotation diversity and intensity, soil and site limitations, residue removal after harvest and other practices such as cover crops, mulching, etc. that are utilized with NT. Key variables that impact NT's effectiveness include the level of soil disturbance that occurs with planting, fertilizer application and other field operation intensity and other cropping practices utilized in conjunction with NT that provide synergistic effects. These include practices such as the use of diverse cropping systems with high residue crops, cover crops grown in the off-season, and other practices/measures that enhance soil health.

The variable effect of NT is especially the case when NT systems include crops with fragile residues, or where disturbance occurs when harvesting crops (e.g. edible beans, potatoes, vegetables, peanuts, cotton, and soybeans). Crops with fragile residues provide less residue cover on the surface, break down quickly, and return less carbon to the soil, which can decrease SOM and increase erosion. When crops with fragile residues are utilized, it is important to utilize additional measures and conservation practices to offset the lack of residue, by adding measures such as cover crops, double cropping, mulch application or rotations with perennial grass or small grains. This is especially important on soils with inherent physical limitations that can reduce yields and plant growth/vigor. These include sandy soils with low available water

Table 1 Impact of r o-till and associated practices.

Impact of NT and Associated Practices (Negative Effects Underlined)	No-till and Associated Practice with the Most Beneficial Impact or Least Detrimental Impact on Soil Physical Properties	Systems Ccompared	Years of NT	Moisture Regions	Soil Temp. Regime	Soil Texture/ Limitation; Location of Study	Reference (First Author and Date)
Improved Ksat	Perennial grasses in rotation (328) and NT best (of cropping systems)	CRP, NT corn–soybean–wheat/clover cover crop, and MT corn–soybean and NT perennial hay/corn	Short term	SH	M	Clay pan soil; Missouri	Jiang et al. 2007
Higher BD, lower POR, higher microporosity, higher AWC	NT with corn–peanuts–cotton–sorghum and soybeans	NT, CT, organic crop production and other land uses	8 years	H	T	Sandy loam and loamy sand soils; Goldsboro, NC	Raczkowski et al. 2012
Slight improvement of SOM	Small grain without fallow (328) and NT best	NT, MT and CT small grain with and without fallow	11 years	SA	F	Coarse sandy soils; SW Saskatchewan	Campbell et al. 1996
Improved SOM, POR, AWC, BD & PR	NT corn–soybean rotation with winter cover crops Hairy Vetch & Cereal Rye (340) best	All NT with corn–soybean rotation with winter cover crops	Short term	H	M	Silt Loam soils; Illinois	Villamil et al. 2006
Impoved SOM & overall physical properties	Cover crops (340) and NT best	NT with various cover crops and fallow	Short term	H	T	Sandy soils; coastal plains of Georgia	Hubbard et al. 2013
Maintained SOM & overall physical properties (with stover removal & NT)	NT without stover removal following harvest is best, long-term tillage and stover removal (328 & 511) worst (stover harvest was sustainable with NT)	NT, CT with and without stover harvest	Long term	H	M	Silt loam soils; Chazy, NY	Moebius-Clune et al. 2008

(*Continued*)

Table 1 (*Continued*)

Impact of NT and Associated Practices (Negative Effects Underlined)	No-till and Associated Practice with the Most Beneficial Impact or Least Detrimental Impact on Soil Physical Properties	Systems Ccompared	Years of NT	Moisture Regions	Soil Temp. Regime	Soil Texture/ Limitation; Location of Study	Reference (First Author and Date)
Ksat, Infilt, AgS, BD, compaction, PR all improved	NT with residue mulch (484) best	NT, RT and CT with and without residue mulch application	Long term	H	M	Silt loam soils; Central Ohio	Kahlon et al. 2013
AgS/size, and SOM and overall physical properties significantly improved	NT with corn–soybean–wheat/cowpea rotation (328) performed best over time	NT and CT corn–soybean, continuous corn, corn–soybean–wheat–cowpea	Long term	H	M	Silt loam soils; South Central Ohio	Aziz et al. 2013
BD unaffected, SOM levels maintained with NT (delined with CT)	NT best especially last 2 years wheat and barley rotation following pasture	NT and CT with wheat & barley rotations following short term pasture wheat, barley, linseed, & peas	9 years	SA/SH	–	Well drained silt loam soils; New Zealand	Francis and Knight 1993
<u>Increased BD</u> and slightly higher SOM in surface only	NT wheat and barley rotation continuous cropping	Same study as above except continuous cropping	9 years	SA/SH	–	Poorly drained silt loam soils; New Zealand	Francis and Knight 1993
Significant increase SOM, POR & AWC	7-year NT best no rotation differences noted	NT and CT with continuous wheat, wheat fava bean and wheat–sulla rotation	3 & 7 years	SH	T	Clay loam soils; Tunisia	Jemai et al. 2013
Significant increase in POR, infil., Ksat, SOM, AgS improved BD	NT with soybean/oat double cropping	NT and CT with soybean/oat	14 years	SH	T	Weakly structured silt loam soils; Northwest Australia	So et al. 2009

BD and PR increased, AgS improved	NT no rotation/cover crop differences with or without grazing (slightly higher PR with cover crop grazing)	NT and CT with summer grain/ winter cover crop and winter grain/ summer cover crop with and without grazing	2.5 years	H	T	Sandy loam surface with clay subsoil (Cecil series); Piedmont Uplands Georgia	Franzluebbers and Stuedemann 2008
Improved AgS except at AL site; higher SOM in surface at all sites nearly three times higher than CT at GA site, BD increased on all sites other properties had no significant change	NT with continuous soybeans in three locations and continuous grain sorghum in GA site	NT and CT with monoculture of Soybeans at locations other than Watkinsville, GA, where monoculture of grain sorghum was used. No sites had cover crops used	5 years Al & MS, 9 years TN, and 15 years in GA	H	T	Loamy sand – Auburn, AL; Sandy clay loam – Watkinsville, GA; Silty clay loam – Verona, MS; Silt loam – Jackson, TN.	Rhoton et al. 1993
Increased macropore connectivity in all studies, inconsistant on all other physical properties	NT but no information shown on crop rotations, cover crops and other management measures	NT only but no information shown on crop rotations cover crops and other management measures (29 studies cited)	No data shown	Varies	Varies	Varies	Strudley et al. 2008
Increased infiltration, AWC	NT with controlled traffic the best	NT with & without controlled traffic, MT with & without controlled traffic	Model	Model	Model	Model	Li et al. 2008
Increased infiltration, AWC and improved BD and PR	NT with winter wheat the best	NT, MT, CT with winter wheat	8 years	SA-SH	T	Silt loam soils; El Reno, Oklahoma	Dao 1993
Ksat decrease, BD increased, macropore connectivity and structure improved	NT with winter wheat	NT, RT, CT with winter wheat	3 years	H	M	Silt loam soils; Raasdorf, Lower Austria	Schwen et al. 2011

(Continued)

Table 1 (*Continued*)

Impact of NT and Associated Practices (Negative Effects Underlined)	No-till and Associated Practice with the Most Beneficial Impact or Least Detrimental Impact on Soil Physical Properties	Systems Ccompared	Years of NT	Moisture Regions	Soil Temp. Regime	Soil Texture/ Limitation; Location of Study	Reference (First Author and Date)
Reduced BD, increased infiltration, increased SOM	NT with summer/winter double cropping the best	(1) long-term (CT) with winter wheat barley and rye (2) long-term no tillage (NT) with summer/winter double cropping (i.e. sorghum cotton or soybean) with winter wheat, rye, barley, or crimson clover	25 years	H	T	Sandy loam surface with clay subsoil (Cecil series); Piedmont Uplands ARS research farm Watkinsville, GA	Franzluebbers 2002
Increased infiltration and SOM	NT with summer/winter double cropping (328) the best	Long-term no tillage (NT) with summer/winter double cropping (i.e. sorghum cotton or soybean) with winter wheat, rye, barley or crimson clover	28 years	H	T	Same study as above	Campbell 2005
Increased Infiltration, other physical properties minimal change	NT wheat/soybeans–corn and wheat/soybeans best cropland (328)	NT with continuous corn, continuous soybeans, wheat/soybean double crop, wheat/soybean double crop-corn, pasture, and tilled fallow	Short term	SH	T	Silt loam soils; Argentina	Sasal et al. 2010
Increased Ksat, improved soil structure	NT with highest level of mulching (484) best	NT with straw after 100s of years of CT	4 years	SA, SH	M, F	Silt loam soils; Northwestern Loess Plateau, China	Zhang et al. 2008
No change in physical properties	NT was equal to MT with Corn Soybean rotation (328) (CT with fallow was worst)	NT, MT, CT with corn soybeans and fallow systems following CT	13 years	SH, H	M, T	Clay pan soil; Kingdom City, Missouri	Blanco-Canqui et al. 2004

Ksat, POR improved	NT with highest level of mulching (484) best	NT with three levels of straw mulch	10 years	H	M	Silt loam soils; Central Ohio	Blanco-Canqui and Lal 2007
Ksat, BD, PR, structure, SOM strongly improved	NT with highest level of mulching (484) best	NT, RT, CT with three levels of mulch	20 years	H	M	Silt loam soils; Central Ohio	Kahlon et al. 2013
Minimal improvements in Ksat SOM, BD, & POR. (long- and short-term NT similar with corn–soybean rotation)	NT with corn soybean rotation was only slightly better than CT (both long and short term). Short term NT similar to long term no till. Sod best with NT (surface only)	NT and CT corn–soybeans; Short term NT after long-term CT; CT after long-term NT; CT after long-term sod; long-term sod.	Long term and short term (4 years)	H	M	Clay loam soils; Essex County, Ontario, Canada	Reynolds et al. 2007
Improved Ksat, POR, SOM; <u>reduced infiltration rate</u>	NT with continuous barley	NT and CT with continuous barley	20 years	SA	Subarctic	Silt loam soils; interior Alaska	Sharratt et al. 2006
<u>BD increased and Ksat and infiltration decreased</u>	The use of fallow in cropping systems (328) had a more negative impact than tillage system (MT vs. NT)	NT and MT with wheat fallow (WF) and wheat sorghum fallow (WSF) rotations	12 years	A, SA	T	Clay loam soils; Bushland, TX	Evett et al. 1999
<u>BD increased, POR decreased</u>, AgS improved	NT corn–soybean rotation (cover crop improved BD and POR compared to no cover crop)	NT and CT corn–soybean rotation with no cover crop, early terminated and late terminated cover crop	3 yeas	H	M	Silt loam soils; Southeast Purdue Agricultural Center Indiana	Stipesevic and Kladivko 2002
<u>Slight decrease in SOM</u>	NT corn–soybean rotation maintained the most SOM	NT, MT, and CT corn–soybean rotation with no cover crops	24 years	H	M	Silt loam fragipan soils; Southern Illinois	Olson et al. 2013

NT, no-till; MT, mulch till; CT, conventional tillage; RT, ridge till; POR, porosity; BD, bulk density; PR, penetration resistance; CRP, Conservation Reserve Program; SOM, soil organic matter; AgS, aggregate stability; Ksat, saturated hydraulic conductivity; AWC, available water capacity; H, humid; SH, sub-humid; SA, semi-arid; A, arid; T, thermic; M, mesic.

Source: Adapted from: USDA-NRCS, 2015 (Kucera et al.).

capacity and cation exchange capacity; soils very high in clay content; soils with root restrictions such as clay pan or fragipans; and soils with other physical limitations that negatively impact plant available water, infiltration and nutrient availability.

Key soil properties discussed in this section are:

- Bulk density;
- Pore size distribution;
- Saturated hydraulic conductivity;
- Available water holding capacity;
- Compaction, penetration resistance, crusting/sealing and infiltration;
- Aggregate size, stability and soil structure; and
- Soil organic matter.

All of these soil properties affect soil health, including vulnerability to water erosion in particular.

2.1 Bulk density

There is an established link between soil BD and erosion, suggesting higher BD generally increases erosion risk (Jepson et al., 1997; Heike and Schmidt, 2013). Reduction in BD occurs immediately following tillage, but this effect reverses over time. The collapse of soil pore space and disrupted activity by microorganisms, microfauna and mesofauna (which contribute to soil physical and chemical structure) results in eventual loss of soil structure and aggregate stability. Conventional tillage such as plowing or tillage systems that leave little protective residue cover can therefore lead to increased erosion and compaction, reduced infiltration and increased penetration resistance compared to NT (Table 1).

In most studies, BD will typically increase in the early stages of NT compared to conventional tillage (Stipesevic and Kladivko, 2002; Schwen et al., 2011). In some long-term NT studies (>12 years), BD may increase compared to the initial BD of the soil prior to starting a continuous NT system (Evett et al., 1999; Blanco-Canqui et al., 2004 Rhoton et al., 1993; Franzluebbers et al., 2008). However, the impact of NT on BD is highly variable depending on factors such as climate, soil type, crop rotation and other practices used in conjunction with NT. The amount of biomass and organic material left on the soil surface and returned to the soil has a dramatic effect on whether BD will decrease or increase when comparing NT to conventional till.

Soils with increased SOM and improved porosity will tend to have a lower BD. Kahlon et al. (2013) and Franzluebbers (2002) are examples of long-term NT studies that show a decrease in BD when NT is implemented with additional measures such as double cropping or application of residue mulch. Under ideal conditions, BD can be reduced even with short-term NT when additional practices such as cover crops are utilized alongside a NT cropping system (Villamil et al., 2006). NT systems in thermic soil temperature regimes tend to have higher BD than mesic regimes, even after long-term NT, unless additional measures are used such as cover crops, mulching, double cropping and use of perennial crops or small grains in rotations (Evett et al., 1999). Climates that are warm and moist will result in higher soil microbial activity. Under these conditions, residue breakdown and SOM mineralization accelerates, leading to lower SOM levels which tend to increase BD.

2.2 Pore size distribution

Characteristics such as soil water retention and pore size distribution have been related to improved soil physical structure and reduced water erosion risk (Gao et al., 2019). Macropores are typically associated with channels created by roots or earthworms. Full width tillage will destroy these channels and reduce macropore numbers and connectivity. The overall trend in the studies reviewed in Table 1 was that macropore connectivity increased under NT, allowing for improved water infiltration. There were inconsistent responses in total porosity and soil BD compared with conventional tillage practices (Stubbs et al., 2004). Porosity initially tends to be lower under NT than in Con T but improves as SOM and aggregate stability improve over time. Increases in porosity increase BD. Jian et al. (2007) found that micropores (<10 μm) increased under NT when compared to Con T, even in root restrictive clay pan soils. When converting from Con T to NT, BD initially increases and then decreases above sea level SOM improves. Including additional measures such as residue mulch application, winter cover crops with fibrous rooting systems, and addition of small grains and other high residue crops in rotation will increase porosity compared to NT systems where they are not utilized (Villamil et al., 2006; Blanco-Canqui et al., 2007; Jemai et al., 2013). Raczkowski et al. (2012) conducted an 8-year NT study in North Carolina (USA) on sandy soils which included rotations with both low residue crops (such as cotton, soybeans and peanuts) and high residue crops (such as corn, small grains and sorghum). The use of low residue crops resulted in higher BD, lower macroporosity and higher microporosity than Con T. Porosity and BD were improved when measures such as high residue crops, cover crops or double cropping, and residue mulch application are utilized along with NT. These factors also help to explain why earthworm numbers tend to increase in NT when compared to full-width tillage systems (Francis et al., 1993; Karlen et al., 1994; Blanco-Canqui and Lal, 2007, 2008).

2.3 Saturated hydraulic conductivity

The overall trend in nearly all studies reviewed in Table 1 was that saturated hydraulic conductivity (typically shortened to Ksat – where K denotes hydraulic conductivity) improved under NT compared to MT and Con T. NT allowed for improved water movement, similar to infiltration measurements. Evett et al. (1999) was the only US study shown in Table 1 where Ksat decreased, but this was associated with the use of fallow in the cropping systems. This highlighted the importance of growing high residue crops and eliminating fallow where possible. Tillage has a pronounced initial effect on soil hydraulic properties, but these effects diminish rapidly (Strudley et al., 2008). The general trend is for NT to provide an increase in saturated hydraulic conductivity (Strudley et al., 2008). NT also increases macropore connectivity, while MT or Con T disrupt this connectivity which negatively impacts Ksat and infiltration.

2.4 Available water holding capacity

The available water holding capacity (AWC) of soils has been related to increased SOM, soil structure, BD, porosity and infiltration, properties which can then reduce water erosion risk, in particular (Smith, 2018). Several studies show that AWC improves when other soil physical properties such as SOM, infiltration rates, BD, soil structure and penetration resistance improve (Dao, 1993; Li et al., 2008; Villamil et al., 2006).

Hudson (1994) evaluated the impact of SOM in surface horizons on AWC in sand, silt loam and silty clay loam soils. Highly positive correlations were found between SOM content and AWC. In all texture groups, as SOM content increased from 0.5% to 3%, AWC of the surface soil more than doubled. In general, long-term NT, along with high residue crops grown each growing season, will enhance physical soil properties through the development and maintenance of water-stable aggregates. Water-stable aggregates are responsible for increasing infiltration and AWC (Brock, 1999). Depending on site and soil conditions, when additional measures/practices such as rotations with small grains, cover crops, mulching or other soil improving practices were implemented along with NT, AWC improved the most. NT systems also had higher levels of soil moisture than Con T where soil moisture levels were monitored (Table 1). The overall trend for NT is to increase macropore connectivity while generating inconsistent responses in total porosity and BD compared with conventional tillage practices. This corresponds to a general increase in ponded or near-zero tension infiltration rates and saturated hydraulic conductivity (Stubbs et al., 2004).

2.5 Compaction, penetration resistance, crusting/sealing and infiltration

Brock (1999) found that long-term NT reduces crusting/sealing, increases infiltration and significantly reduces runoff, even in warm moist climates in the Southeastern US. This reduces, in most studies where compaction, penetration resistance, crusting/sealing and infiltration were analyzed, these properties were improved with NT compared to Con T (Table 1) (Dao, 1993; Franzluebbers, 2002; Villamil et al., 2006; Li et al., 2008; Sasal et al., 2010; So et al., 2009). Sharratt et al. (2006) was the only study where NT did not have better results compared to conventional tillage on these properties. In this study, infiltration was less with NT than it was with a fall chisel plow system for continuous small grains cultivation in Alaska. This was attributed to extreme conditions and the limited time where soils are not frozen. Mesic and frigid soil temperature regimes have an advantage over thermic temperature regimes because of freeze-thaw cycles that can reduce surface compaction and penetration resistance. Jabro (2012), for example, found that frequent freeze-thaw cycles over the winter reduced soil penetration resistance by 73% and 68% over the winter at depths of 0–10 and 10–20 cm, respectively. In unfrozen compacted soils, penetration resistance was also reduced by 50% and 60% over winter at 0–10 and 10–20 cm depths.

2.6 Aggregate size, stability and soil structure

Aggregate stability and size are very important for retaining soil structure and thus reducing erosion. In studies where aggregate size, stability and structure were analyzed, these properties were improved by NT compared to conventional tillage (Table 1). As an example, improvement in aggregate stability occurred even after only 3 years of NT in sandy Piedmont soils in Georgia (USA) in thermic soils (Franzluebbers et al., 2008). Brock (1999) reported similar findings for long-term NT but that water stable aggregates cannot be sustained through Con T since there is not sufficient residue cover to protect against surface crusting. This emphasizes the importance of limiting soil disturbance to improve aggregation and structure, especially in structurally weaker soils such as clay pan soils or sandy (Blanco-Canqui et al., 2004). Karlen et al. (1994) found a significant increase in aggregate size following 10 years of NT corn on highly erodible silt loam soils on 10–13% slopes near Lancaster (Wisconsin, USA). Rhoton et al. (1993) conducted a 15-year study on four Southeastern US soils of differing textures and found in all cases that aggregate stability was higher under NT than Con T. These and other studies indicate that these physical properties are improved with NT regardless of temperature and moisture region throughout the US, when compared to full-width tillage systems.

2.7 Soil organic matter

As indicated by the studies in Table 1, the extent of SOM improvement in soils is not solely dependent NT. Additional practices such as mulching, crop rotation, forage or biomass crops and cover crops can all have an impact on SOM levels. In many long-term studies, where low residue crops were used in rotation, SOM did not improve, but losses were minimized by utilizing NT. This can be seen for example in a 24-year-study NT with a corn–soybean rotation on eroded fragipan soils in Southern Illinois NT plots when compared to Con T (Olson, 2013). Even though erosion was lower and SOM higher on the NT plots, SOM did not increase over the 24 years of the study. Other studies also suggest that gains in SOM were not likely without additional measures, especially in thermic soil temperature regimes (Rhoton et al., 1993). In these situations, it is necessary to increase the use of high residue crops, perennial legumes, small grains or high carbon cover crops such as cereal rye to increase or maintain SOM levels. Aziz et al. (2013) demonstrated that SOM could be significantly increased in silt loam soils by adding crop diversity to traditional corn–soybean rotations or other rotations that utilize a significant amount of low residue crops. Other long-term NT studies in thermic soil temperature regimes demonstrate that addition of mulching, high residue crops, cover crops and diverse rotations can increase SOM (Kahlon et al., 2013; So et al., 2009; Sharratt, 2006; Rhoton et al., 1993; Aziz et al., 2013).

3 No-till practices and water erosion risk

Soil erosion by water is a function of climatic and topographic factors, soil characteristics, structural measures and cropping practices (see Chapter 2 by Dennis Flanagan in this book). When soil is tilled and loosened, and residue is buried and crops planted, the potential for accelerated soil loss from water erosion increases. Annual soil loss (A) by water erosion from various slopes and slope lengths (LS), soils (K), climatic zones (*R*), crops and production systems (C), and conservation practices (P) are predicted through application of the Universal Soil Loss Equation (USLE) developed by Wischmeier and Smith (Wischmeier and Smith, 1965, 1978). USLE was developed from 10 000 plot years of soil loss data from various locations, soils and climatic zones, mostly collected before NT practices were common. Successive iterations include Revised USLE (RUSLE), Water Erosion Prediction Program (WEPP) and now RUSLE2 (USDA, 2013).

As more soil loss data from NT cropland sites became available, the original USLE equations were expanded to include the effects of this production system. Initially benefits of NT with cover were predicted to provide a 5× to 10× reduction in water erosion risk when compared with moldboard plowing.

The early assessment of NT reduction in water erosion was based on a limited data set with some studies suggesting these were too conservative in assessing erosion reduction benefits. Harrold and Edwards (1974) analyzed data from the two watersheds for seven large storms during the period of 1941–1969 and found similar amounts of soil loss. However, after adopting NT in one watershed and continuing the other in conventional tillage, the cumulative soil loss differences in the years from 1970 to 1973 clearly demonstrated a higher than predicted effectiveness of NT in controlling erosion. Van Doren and Triplett (1982) e.g. used USLE to estimate land-use potential in Ohio with various tillage practices, including accounting for soil loss. The best NT system achieved a score of 84%, significantly ahead of conventional tillage, but this increased to 90–95% with revised C factors in an updated version of USLE. In a subsequent study, Van Doren et al. (1984) found that soil loss for sites not tilled for more than 1 year was 90% less than predicted from USLE values at that time. Measurements in 2004, on NT plots continuously maintained for 42 year under corn in Wooster (Ohio, USA), on uniform slopes of 2.5–4.0%, showed an elevation difference of 9.0 cm between NT plots and plow tillage (PT) plots (unpublished data). Assuming a BD of 1.40 Mg ha^{-1}, approximately 1260 Mg more soil was lost from the PT plots compared with the NT plots for an average of 30 Mg ha^{-1} $year^{-1}$.

Several mechanisms act to reduce water erosion with NT-based crop production. Raindrop impact provides the major energy source in dislodging soil particles to initiate erosion. Residue cover on the soil surface intercepts raindrops and absorbs much of this energy (Bradford and Huang, 1994). In addition, untilled soil requires more energy to dislodge particles than for bare soil loosened by tillage (Wilson et al., 2008). After particles are dislodged, they are transported by overland flow of runoff. Transport is necessary for erosion to occur. Residue, applied mulch and vegetative cover from standing plants slow runoff water flow rates, reducing the suspension and transport of soil particles. Plant roots growing at or near the soil-mulch also provide an interface to bind soil particles and create additional resistance to dislodging particles and resulting soil movement.

The effectiveness of cover is based on factors such as crop residue cover, canopy cover and perennial and annual cover (Panagos et al., 2015). The effects of crop canopy and surface residue cover are not additive because canopy cover affects raindrop impact, but surface residue also reduces runoff velocity, raindrop impact and surface crusting. Vegetative cover and standing residue have the additional benefit of being anchored in the soil surface and will not be displaced by run off water.

WEPP (USDA, 2024) and RUSLE2 (USDA, 2013) water erosion models were utilized by the USDA in 2020 to evaluate the impact of tillage on predicted water erosion rates (Wells et al., 2024). Forty-two years of precipitation data from a National Weather Service station in Hoopeston (Vermilion County in Illinois, USA) were used to provide climate inputs to both RUSLE2 and WEPP. Tillage systems evaluated included fallow, corn-soybean rotations with NT, conservation tillage and conventional tillage with and without fall cereal rye cover crop management systems had a large effect on the predicted soil loss by both models (Fig. 4).

The USDA found that tilled fallow management (where no crops are grown and the soil is continually disturbed) had the greatest predicted soil loss, with both WEPP and RUSLE2 average annual soil loss values around 30 T/ac-year (Table 2). A conventional tillage corn-soybean rotation dramatically decreased predicted soil loss levels to about 8 T/ac-year. Use of a fall rye cover crop after soybeans in the rotation decreased average soil loss to about 6 T/ac-year. The CT corn-soybean rotation had average soil loss levels at about 5 T/ac-year. The use of a fall rye cover crop after soybeans decreased average predicted soil loss for both models to about 3 T/ac-year. NT corn-soybean management both without and with a fall cover crop had the lowest predicted soil loss levels of about 1 T/ac-year for both models (Table 2, Fig. 4).

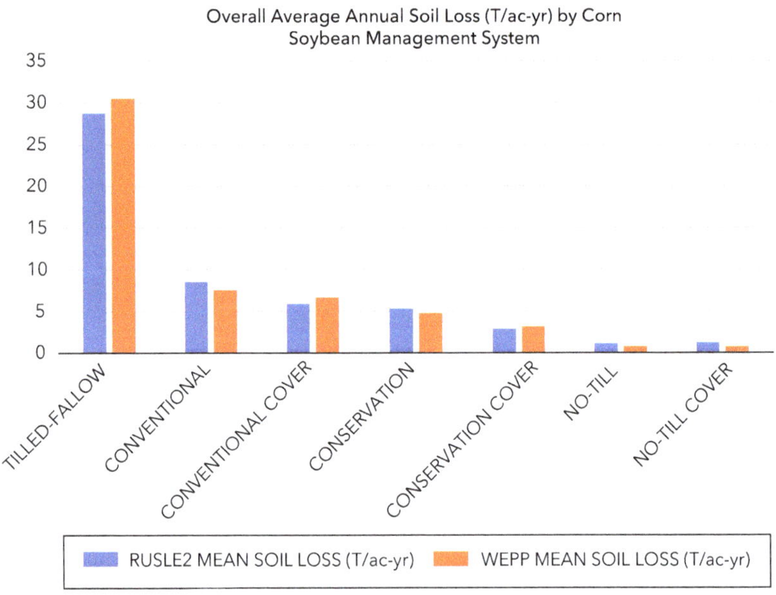

Figure 4 RUSLE2 and WEPP predicted average annual soil loss at Hoopeston, Vermilion County, Illinois, by management system across all soil types, slope gradients, slope lengths and crop yield levels. Conservation = conservation mulch tillage system; conventional = conventional tillage; cover = fall cereal rye cover crop.

Table 2 Average and median RUSLE2 and WEPP predicted soil loss on tilled fallow and corn–soybean management system across all model simulations at Hoopeston, Vermilion County, Illinois

Management	RUSLE2 Mean Soil Loss (T/ac-year)	WEPP Mean Soil Loss (T/ac-year)	RUSLE2 STDEV (T/ac-year)	WEPP STDEV (T/ac-year)	RUSLE2 Median Soil Loss (T/ac-year)	WEPP Median Soil Loss (T/ac-year)
Tilled fallow	28.74	30.51	22.27	27.91	24.99	22.15
Conventional	8.56	7.56	6.41	6.70	7.53	5.59
Conventional cover	5.89	6.64	4.33	5.91	5.20	4.84
Conservation	5.26	4.73	3.86	4.35	4.64	3.30
Conservation cover	2.86	3.11	1.98	2.88	2.59	2.08
No-till	1.11	0.74	0.80	0.29	0.95	0.71
No-till cover	1.22	0.72	0.81	0.27	1.11	0.70

4 No-till practices and wind erosion risk

Though less well studied than water erosion, wind erosion is still a major cause of soil loss (Bartkowski et al., 2023). The energy source that causes wind erosion is wind velocity at the soil surface (Chepil and Woodruff, 1963). Sand grains set in motion by wind move across the soil surface, dislodging other particles in a process called creep and saltation (Fig. 5). Loosened silt and clay particles can also be suspended and carried in the wind stream and remain aloft until wind velocity decreases then are deposited (Fig. 5) (Duniway et al., 2019). Detached soil particles are subject to transport by the wind, either through the air or along the surface when the soil surface is left unprotected by residue or canopy cover. When the soil surface is left unprotected by residue or canopy cover, the distance, height and duration of transport are dependent largely on the wind speed. During the 1930s wind erosion in the great plains of the United States created severe dust storms in the so-called 'Dust Bowl' with disastrous consequences for agriculture and the environment across the southern plains and beyond.

When adequate crop residues, especially standing stubble, are left on the soil surface, as is the case with NT, this residue cover reduces wind energy and protects the soil surface reducing soil loss to negligible levels. Maintaining crop residues is thus the single most practical way to reduce wind erosion (Fryrear, 1985). NT systems that provide residue cover throughout the year, especially standing residue, are most impactful in preventing wind erosion. When protective residue cover is left on the soil surface sediment transport of detached soil particles by wind transportation processes shown below can be addressed for example by:

- Reducing the wind force at the soil surface (process 1 shown in Fig. 6);
- Creating a soil surface more resistant to wind forces (process 2 shown in Fig. 6); and
- Trapping saltating particles to reduce the abrasion of soil surfaces downwind (process 3 shown in Fig. 6).

The cardinal rule of wind erosion control is to keep the land continuously covered with vegetation or crop residue. The following are paramount as alternatives are considered (Nordstrum and Hatta, 2004; Tatarko et al., 2019):

- Return all land unsuited to cultivation to permanent cover;
- Maintain maximum possible cover on the surface during wind erosion periods;
- Always maintain stable field borders or boundaries; and
- Keep all residue standing as long as possible.

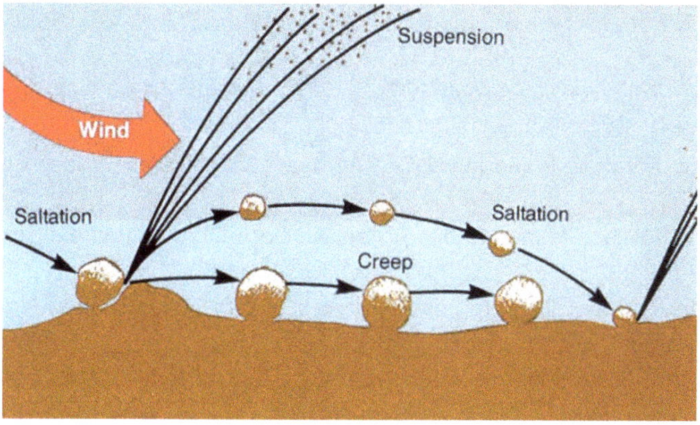

Figure 5 Wind erosion transport processes.

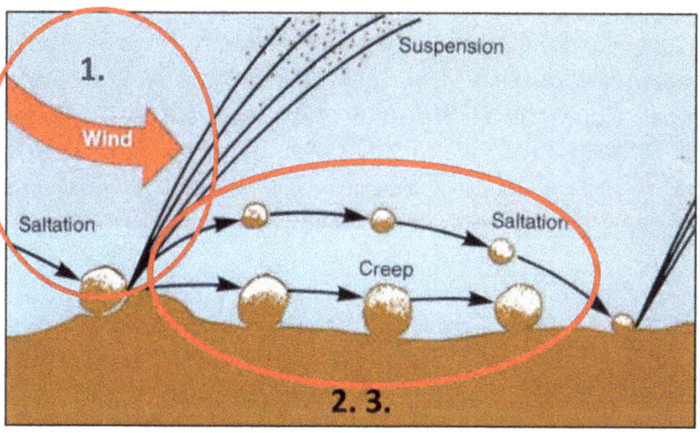

Figure 6 Wind erosion transport processes and three processes that trigger wind erosion.

Standing residue is at least three times more effective at controlling wind erosion than flat residue.

The Wind Erosion Prediction System (WEPS) is a process-based, daily time-step model that simulates weather, field condtions and estimates average annual wind erosion and for each crop interval period (USDA, 2023). In this chapter, WEPS Version 1.5 Wind Erosion Prediction System (usda.gov) (for use by USDA-NRCS and Technical Service Providers) was utilized to run six crop rotation/residue management systems.

Six rainfed management systems on a square 160 acre tract of land were evaluated in Kimball County Nebraska as follows:

- Crop Rotation Management System 1 - Wheat-Fallow with Spring Mulch Tillage
- Crop Rotation Management System 2 - Wheat-Fallow Fall, Spring and Summer Mulch Tillage
- Crop Rotation Management System 3 - Corn-Millet-Wheat Continous No-Till
- Crop Rotation Management System 4 - Corn-Drybean-Wheat Continuous No-Till
- Crop Rotation Management System 5 - Corn-Fallow-Wheat No-Till and Mulch Till
- Crop Rotation Management System 6 - Corn-Fallow-Wheat Eco-fallow/ Continous No-Till

The average annual erosion rates of the six management systems varied from 0.0 tons with a continuous NT system to as high as 46.2 tons (41.9 metric tons/ acre with a conventional tillage wheat fallow system (Table 3)). The importance of maintaining residue cover, especially NT, is demonstrated by these examples from Kimball County in far Western Nebraska near the border with both

Table 3 Average annual, maximum and minimum erosion rates for five management systems (US tons and metric tons equivalent above 1 US ton)

Crop Rotation/Management	Average Annual Wind Erosion Rate for Entire Crop Rotation	Maximum Erosion Rate/ Rotation Period	Minimum Erosion Rate/Rotation Period
System 1. Wheat–fallow spring mulch tillage	3.0 tons (2.7 metric tons)/ acre	6.0 tons (5.4 metric tons)/acre fallow year 1	0.0 tons/acre winter wheat year 2
System 2. Wheat–fallow fall, spring and summer Mulch Tillage	42.7 tons (38.7 metric tons)/ acre	64.3 tons (58.3 metric tons)/acre fallow year 1	21.0 tons (19 metric tons)/acre winter wheat year 2
System 3. Corn–millet–wheat continuous NT	0.0 tons/acre	0 tons/acre (all 3 years)	0.0 tons/acre (all 3years)
System 4. Corn–dry beans–wheat continuous NT	Trace	0.1 tons/acre dry beans year 2	0.0 tons/ac corn year 1 and winter wheat year 3
System 5. Crop rotation management corn–fallow–no-till and mulch till during fallow year	2.0 tons (1.8 metric tons)/ acre	6.1 tons (5.5 metric tons)/acre fallow year 2	0.0 tons/acre corn year 1 and winter wheat year 3
System 6. Corn–fallow–wheat eco-fallow/continuous no-till	0.1 Tons/acre	0.3 Tons/acre fallow year 2	0.0 tons/acre corn year 1 and einter eheat yar 3

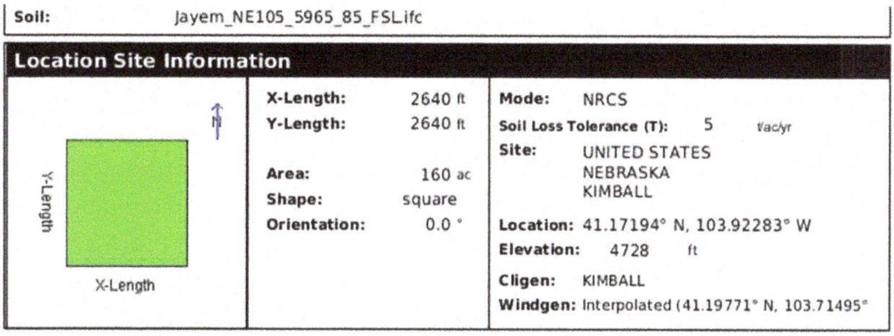

Figure 7 Field configuration, location and soil map unit used for WEPS simulation runs.

Colorado and Wyoming (Fig. 7). A Jayem Fine Sandy Loam soil map unit on a square 160 acre field was used in this simulation (Fig. 7). The top 150 mm on the soil surface contains 67.6% sand, 20.4% silt and 12.0% clay. This soil map unit component is considered highly erodible and prone to wind erosion without appropriate conservation measures. Crop rotation sequence, tillage and planting operations, dates, erosion rate details and graphs showing wind erosion timing detailing wind erosion from creep plus saltation and suspension for each of the six crop rotation management systems are detailed in Figs. 8–13 (see Appendix 1).

5 Conclusion

In order to maximize benefits of NT and CT systems, it is important to follow four key soil health planning principles:

- Minimize soil disturbance, whether that is physical (tillage), chemical (excessive nutrients or pesticides) or biological (creating favorable conditions for diseases or weeds). Any necessary managed disturbance should be kept to a minimum.
- Maximize cover of the soil with living or dead plant material.
- Maximize diversity. This involves introducing greater temporal and spatial diversity with crops. This can include the use of perennial crops, diverse crop rotations and/or cover crops. Adding grazing animals to the cropping system would be another means of increasing diversity in the system.
- Maximize the time and distribution of living roots in the soil. This will largely depend on climate and cropping system. The goal is to have growing plants for as much of the year as practicable to feed soil microbes throughout the year.

When implementing all four principles together, improvements in soil physical, chemical and biological properties can be optimized. A 'quality' NT system needs to account for changes in weed and pest types and pressure, nutrient cycling changes and changes in soil properties. The ideal is to have every seed placed at the same depth with even emergence and to manage residues from harvest to planting in order to create consistent residue coverage and field conditions. The length of time that a NT system is in place is very important since there is typically a transition period before soil health improvements and benefits are realized.

Some practical recommendations for adopting a NT system include starting to plan at least 1 year before NT implementation:

Preparatory measures:

- Ensure a smooth field surface when necessary;
- Ensure sufficient fertility and appropriate soil pH;
- Ensure adequate drainage;
- Deal with existing compaction issues such as plow pans;
- Ensure you have appropriately adapted planting, fertilizer and harvest equipment;
- Scout your fields in the fall and winter prior to starting a NT system for weeds and other problems to properly account for them;
- Ensure you can achieve even distribution of crop residue after harvest; and
- Experiment with cover crops strategically before full NT implementation.

Pick an easy entry point and starting crop:

- After a perennial crop, where soil structure is good and where weed issues are minimized;
- After a cover crop, where adding a cover crop will leave ideal soil properties, minimize weed issues and speed up the process of building soil health; and
- Plant the right crop your first year that is best adapted to a new NT system. For example in the US Midwest, soybeans are a good entry point for No-Till because they are more forgiving of soil conditions and nitrogen immobilization is not an issue.

Select the right seeds:

- Select for more than just yield. Including good seedling and root vigor; and
- Talk to local producers using NT about managing pests through seed treatment.

Set up your planter correctly:

- Every seed should be placed at exactly the same depth at exactly the same spacing in exactly the same environment;
- Planter should be level. Row units run true;
- Depth gauging wheels should have uniform pressure;
- Row cleaners should move residue, not soil. The row cleaner should not work against depth control; and
- Sharp disk openers and quality closing wheels should be properly aligned.

Plant according to soil conditions, not the calendar:

- Stay out of fields that are too wet;
- Follow equipment manufacturer recommendations for planting and sprayer speed; and
- Get the planting pass right. You get one chance to plant that seed correctly.

Seek advice and recommendations from successful no-tillers:

- NT is a completely different system of farming. Managing the system the same way as a conventional system (e.g. in terms of nutrient, pest and residue management) can lead to disappointment.
- Be prepared to learn and adapt.

6 Appendix: Wind Erosion Prediction System (WEPS) results from six crop rotation/residue management systems

Rotation Summary

Date	Operation	Crop
Jun 01, 01	Sweep plow wider than 40 inch, mulch treader	
Aug 30, 01	Rodweeder	
Sep 10, 01	Drill or air seeder, double disk	Wheat, winter, grain
Apr 15, 02	Sprayer, post emergence, fertilizer tank mix	weed residue; 0-3 mo
Jul 10, 02	Harvest, killing crop 50pct standing stubble	Wheat, winter, grain

Erosion

		Gross Loss	Net Soil Loss From Field (t/ac)			
Period	Crop/Residue	t/ac	Total	Creep+Salt	Suspen.	PM10
Rot. year: 1		6.0	6.0	2.3	3.7	0.10
Rot. year: 2	Wheat, winter, grain	0.0	0.0	0.0	0.0	0.00
Ave. Annual		3.0	3.0	1.2	1.8	0.05

Crop Interval Erosion

			Gross Loss	Net Soil Loss From Field (t/ac)			
Date Range	Days	Crop	t/ac	Total	Creep+Salt	Suspen.	PM10
Jul 10, 02 - Jul 10, 02	730	Wheat, winter, grain	6.0	6.0	2.3	3.7	0.10

Harvests

Date	Crop	Residue lb/ac	Harvest Yield	Yield % Moisture
Jul 10, 02	Wheat, winter, grain	4,238	42 bu/ac	11

Wind Erosion Soil Loss (Creep+Salt, Susp)

WheatFallowTillage1

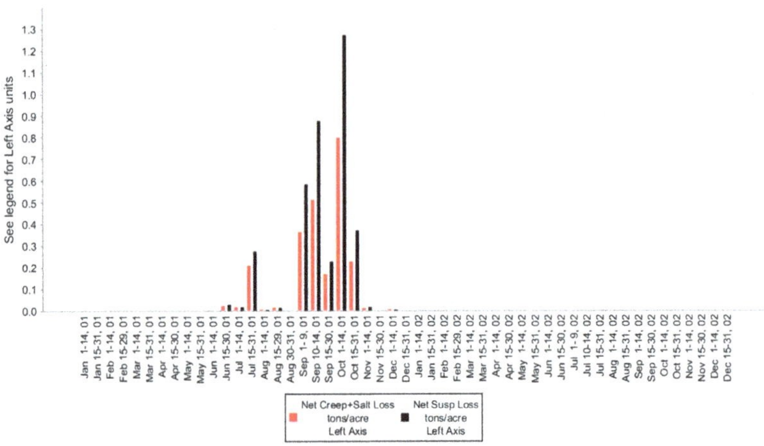

WEPS 1.6.36 Printed Monday, December 11, 2023, 04:27 PM Page 1 of3

Figure 8 Crop rotation management system 1 – Wheat-fallow spring mulch tillage.

Rotation Summary

Date	Operation	Crop
Jun 01, 01	Sweep plow wider than 40 inch, mulch treader	
Aug 30, 01	Rodweeder	
Sep 10, 01	Drill or air seeder, double disk	Wheat, winter, grain
Apr 15, 02	Sprayer, post emergence, fertilizer tank mix	weed residue; 0-3 mo
Jul 10, 02	Harvest, killing crop 50pct standing stubble	Wheat, winter, grain
Aug 10, 02	Disk, tandem heavy primary	
Oct 10, 02	Cultivator, field 6-12 in sweeps, coil tine har	

Erosion

Period	Crop/Residue	Gross Loss t/ac	Net Soil Loss From Field (t/ac) Total	Creep+Salt	Suspen.	PM10
Rot. year: 1		64.3	64.3	22.7	41.6	1.13
Rot. year: 2	Wheat, winter, grain	21.0	21.0	8.0	13.0	0.35
Ave. Annual		42.7	42.7	15.4	27.3	0.74

Crop Interval Erosion

Date Range	Days	Crop	Gross Loss t/ac	Net Soil Loss From Field (t/ac) Total	Creep+Salt	Suspen.	PM10
Jul 10, 02 - Jul 10, 02	730	Wheat, winter, grain	85.4	85.4	30.8	54.6	1.48

Harvests

Date	Crop	Residue lb/ac	Harvest Yield	Yield % Moisture
Jul 10, 02	Wheat, winter, grain	4,161	41 bu/ac	11

Wind Erosion Soil Loss (Creep+Salt, Susp)

NRCS Natural Resources Conservation Service

WheatFallowTillage2_calib

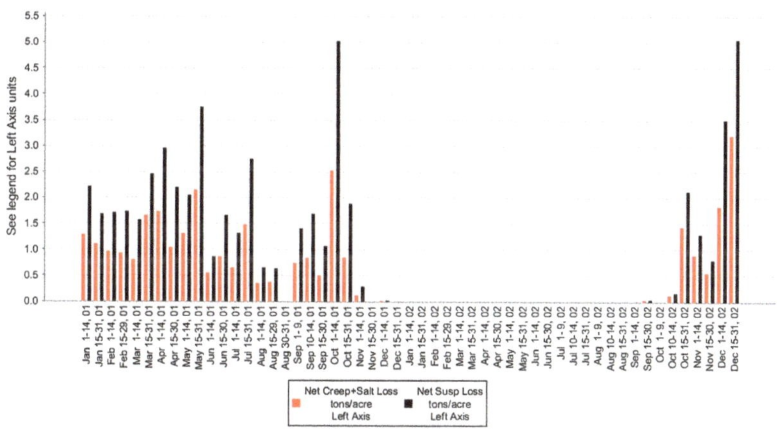

WEPS 1.6.36 Printed Monday, December 11, 2023, 04:34 PM Page 1 of3

Figure 9 Crop rotation management system 2 - wheat-fallow fall, spring and summer mulch tillage.

Rotation Summary

Date	Operation	Crop
May 01, 01	Planter, double disk opnr	Corn, grain, seed
Oct 05, 01	Harvest, killing crop 50pct standing stubble	Corn, grain, seed
Jun 01, 02	Drill or air seeder, double disk	Millet, proso, grain
Sep 01, 02	Harvest, killing crop 50pct standing stubble	Millet, proso, grain
Sep 10, 02	Drill or air seeder, single disk openers 7-10 in spac.	Wheat, winter, grain
Jul 10, 03	Harvest, killing crop 50pct standing stubble	Wheat, winter, grain

Erosion

		Gross Loss	Net Soil Loss From Field (t/ac)			
Period	Crop/Residue	t/ac	Total	Creep+Salt	Suspen.	PM10
Rot. year: 1	Corn, grain, seed	0.0	0.0	0.0	0.0	0.00
Rot. year: 2	Millet, proso, grain	0.0	0.0	0.0	0.0	0.00
Rot. year: 3	Wheat, winter, grain	0.0	0.0	0.0	0.0	0.00
Ave. Annual		0.0	0.0	0.0	0.0	0.00

Crop Interval Erosion

			Gross Loss	Net Soil Loss From Field (t/ac)			
Date Range	Days	Crop	t/ac	Total	Creep+Salt	Suspen.	PM10
Jul 10, 03 - Oct 05, 01	452	Corn, grain, seed	0.0	0.0	0.0	0.0	0.00
Oct 05, 01 - Sep 01, 02	331	Millet, proso, grain	0.0	0.0	0.0	0.0	0.00
Sep 01, 02 - Jul 10, 03	312	Wheat, winter, grain	0.0	0.0	0.0	0.0	0.00

Harvests

Date	Crop	Residue lb/ac	Harvest Yield	Yield % Moisture
Oct 05, 01	Corn, grain, seed	6,309	113 bu/ac	16
Sep 01, 02	Millet, proso, grain	7,830	35 bu/ac	10
Jul 10, 03	Wheat, winter, grain	4,133	41 bu/ac	11

Wind Erosion Soil Loss (Creep+Salt, Susp)

NRCS Natural Resources Conservation Service

corn-millet-wheatNT

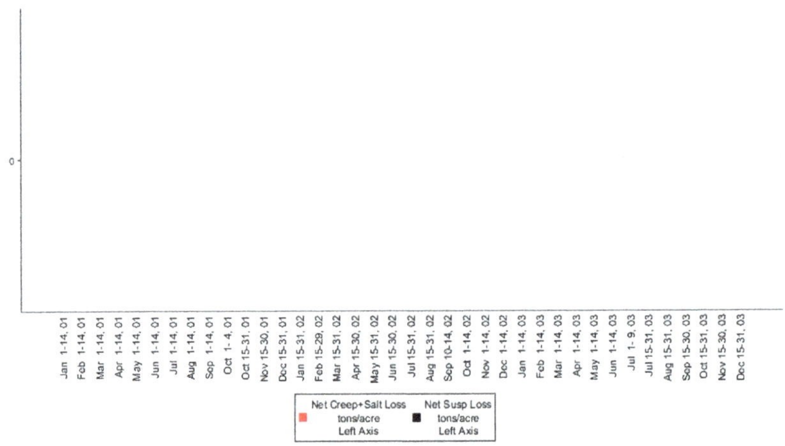

WEPS 1.6.36 Printed Monday, December 11, 2023, 04:48 PM Page 1 of4

Figure 10 Crop rotation management system 3 - corn-millet-wheat continous no-till.

Rotation Summary

Date	Operation	Crop
May 01, 01	Planter, double disk opnr	Corn, grain, seed
Oct 05, 01	Harvest, killing crop 50pct standing stubble	Corn, grain, seed
May 01, 02	Drill or air seeder, double disk	Bean, field, dry
Sep 01, 02	Harvest, killing crop 10pct standing stubble	Bean, field, dry
Sep 20, 02	Drill or air seeder, single disk openers 7-10 in spac.	Wheat, winter, grain
Jul 10, 03	Harvest, killing crop 50pct standing stubble	Wheat, winter, grain

Erosion

		Gross Loss	Net Soil Loss From Field (t/ac)			
Period	Crop/Residue	t/ac	Total	Creep+Salt	Suspen.	PM10
Rot. year: 1	Corn, grain, seed	0.0	0.0	0.0	0.0	0.00
Rot. year: 2	Bean, field, dry	0.1	0.1	Trace	0.1	Trace
Rot. year: 3	Wheat, winter, grain	0.0	0.0	0.0	0.0	0.00
Ave. Annual		Trace	Trace	Trace	Trace	Trace

Crop Interval Erosion

			Gross Loss	Net Soil Loss From Field (t/ac)			
Date Range	Days	Crop	t/ac	Total	Creep+Salt	Suspen.	PM10
Jul 10, 03 - Oct 05, 01	452	Corn, grain, seed	0.0	0.0	0.0	0.0	0.00
Oct 05, 01 - Sep 01, 02	331	Bean, field, dry	0.0	0.0	0.0	0.0	0.00
Sep 01, 02 - Jul 10, 03	312	Wheat, winter, grain	0.1	0.1	0.0	0.1	0.00

Harvests

Date	Crop	Residue lb/ac	Harvest Yield	Yield % Moisture
Oct 05, 01	Corn, grain, seed	4,922	86 bu/ac	16
Sep 01, 02	Bean, field, dry	1,184	1166 lbs/ac	12
Jul 10, 03	Wheat, winter, grain	4,177	41 bu/ac	11

Wind Erosion Soil Loss (Creep+Salt, Susp) NRCS Natural Resources Conservation Service

corn-millet-wheatNT_calib

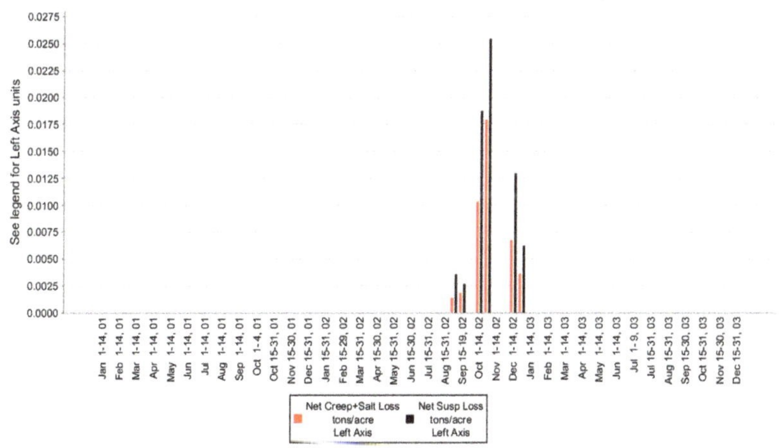

WEPS 1.6.36 Printed Monday, December 11, 2023, 06:46 PM Page 1 of4

Figure 11 Crop Rotation Management System 4 – corn-fallow-wheat continuous no-till.

Rotation Summary

Date	Operation	Crop
May 01, 01	Planter, double disk opnr with starter fertilizer	Corn, grain, seed
Jun 15, 01	Sprayer, post emergence	weed residue; 0-3 mo
Oct 05, 01	Harvest, killing crop 50pct standing stubble	Corn, grain, seed
Nov 01, 01	Chisel, st. pt	
Jul 01, 02	Sweep plow wider than 40 inch, mulch treader	
Aug 01, 02	Sweep plow wider than 40 inch, mulch treader	
Aug 15, 02	Rodweeder	
Sep 10, 02	Drill or air seeder, double disk	Wheat, winter, grain
Apr 01, 03	Sprayer, post emergence, fertilizer tank mix	weed residue; 0-3 mo
Jul 10, 03	Harvest, killing crop 50pct standing stubble	Wheat, winter, grain

Erosion

		Gross Loss	Net Soil Loss From Field (t/ac)			
Period	Crop/Residue	t/ac	Total	Creep+Salt	Suspen.	PM10
Rot. year: 1	Corn, grain, seed	0.0	0.0	0.0	0.0	0.00
Rot. year: 2		6.1	6.1	2.4	3.7	0.10
Rot. year: 3	Wheat, winter, grain	0.0	0.0	0.0	0.0	0.00
Ave. Annual		2.0	2.0	0.8	1.2	0.03

Crop Interval Erosion

			Gross Loss	Net Soil Loss From Field (t/ac)			
Date Range	Days	Crop	t/ac	Total	Creep+Salt	Suspen.	PM10
Jul 10, 03 - Oct 05, 01	452	Corn, grain, seed	0.0	0.0	0.0	0.0	0.00
Oct 05, 01 - Jul 10, 03	643	Wheat, winter, grain	6.1	6.1	2.4	3.7	0.10

Harvests

Date	Crop	Residue lb/ac	Harvest Yield	Yield % Moisture
Oct 05, 01	Corn, grain, seed	6,264	112 bu/ac	16
Jul 10, 03	Wheat, winter, grain	4,193	41 bu/ac	11

Wind Erosion Soil Loss (Creep+Salt, Susp)

NRCS Natural Resources Conservation Service

com-millet-wheatNT_calib_1

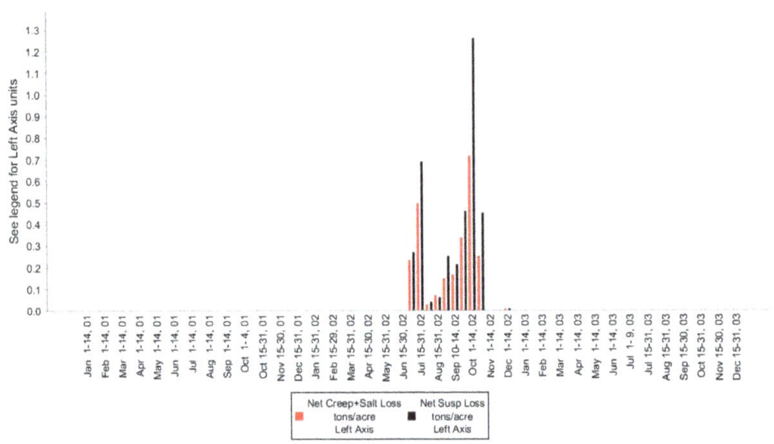

WEPS 1.6.36 Printed Monday, December 11, 2023, 06:56 PM Page 1 of4

Figure 12 Crop rotation management system 5 - corn-fallow-wheat NT and mulch till.

Rotation Summary

Date	Operation	Crop
May 01, 01	Sprayer, pre-emergence	weed residue; 0-3 mo
May 01, 01	Planter, double disk opnr with starter fertilizer	Corn, grain, seed
Jun 15, 01	Sprayer, post emergence	weed residue; 0-3 mo
Oct 05, 01	Harvest, killing crop 50pct standing stubble	Corn, grain, seed
Jun 01, 02	Sprayer, pre-emergence	weed residue; 0-3 mo
Sep 10, 02	Drill or air seeder, double disk	Wheat, winter, grain
Apr 01, 03	Sprayer, post emergence, fertilizer tank mix	weed residue; 0-3 mo
Jul 10, 03	Harvest, killing crop 50pct standing stubble	Wheat, winter, grain

Erosion

		Gross Loss	Net Soil Loss From Field (t/ac)			
Period	Crop/Residue	t/ac	Total	Creep+Salt	Suspen.	PM10
Rot. year: 1	Corn, grain, seed	0.0	0.0	0.0	0.0	0.00
Rot. year: 2		0.3	0.3	0.1	0.2	Trace
Rot. year: 3	Wheat, winter, grain	0.0	0.0	0.0	0.0	0.00
Ave. Annual		0.1	0.1	Trace	0.1	Trace

Crop Interval Erosion

			Gross Loss	Net Soil Loss From Field (t/ac)			
Date Range	Days	Crop	t/ac	Total	Creep+Salt	Suspen.	PM10
Jul 10, 03 - Oct 05, 01	452	Corn, grain, seed	0.0	0.0	0.0	0.0	0.00
Oct 05, 01 - Jul 10, 03	643	Wheat, winter, grain	0.3	0.3	0.1	0.2	0.01

Harvests

Date	Crop	Residue lb/ac	Harvest Yield	Yield % Moisture
Oct 05, 01	Corn, grain, seed	6,284	112 bu/ac	16
Jul 10, 03	Wheat, winter, grain	4,126	41 bu/ac	11

Wind Erosion Soil Loss (Creep+Salt, Susp)

NRCS Natural Resources Conservation Service

corn-millet-wheatNT_calib_2

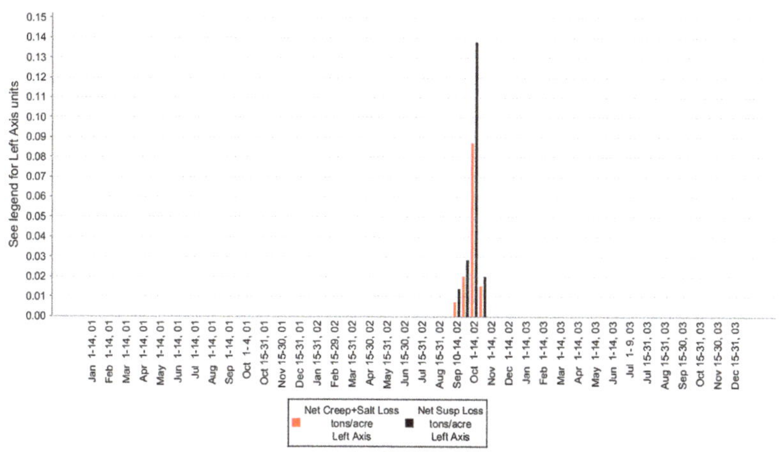

WEPS 1.6.36 Printed Monday, December 11, 2023, 07:17 PM Page 1 of4

Figure 13 Crop rotation management system 6 – corn-fallow-wheat eco-fallow/continous no-till.

7 References

Aziz, I., Mahmood, T. and Islam, K. R. 2013. Effect of long-term no-till and conventional tillage practices on soil quality. *Soil and Tillage Research* 131, 28–35.

Baker, C. J. and Saxton, K. E. (Eds.). 2007. *No-Tillage Seeding in Conservation Agriculture*, 2nd ed. FAO and CAB International.

Bartkowski, B., Schepanski, K., Bredenbeck, S. and Mulier, B. 2023. Wind erosion in European agricultural landscapes: more than physics. *People Nature* 5(1), 34–44.

Blanco-Canqui, H., Gantzer, C. J., Anderson, S. H. and Alberts, E. E. 2004. Tillage and crop influences on physical properties for an epiaqualf. *Soil Science Society of America Journal* 68, 567-576.

Blanco-Canqui, H. and Lal, R. 2007. Impacts of long-term wheat straw management on soil hydraulic properties under no-tillage. *Soil Science Society of America Journal* 71, 1166–1173.

Blanco-Canqui, H. and Lal, R. 2008. Axle-load impacts on hydraulic properties and corn yield in no-till clay and silt loam. *Agronomy Journal* 100, 1673–1680.

Bradford, J. and Huang, C.-H. 1994. Interill soil erosion as affected by tillage and soil cover. *Soil and Tillage Research* 31(4), 353-361.

Brock, B. G. 1999. Rx for soil quality = long-term no-till. In R. Lal (Ed.), *Soil Quality and Soil Erosion, Chapter 10*. Ankeny, IA: Soil and Water Conservation Society.

Campbell, C. A., McConkey, B. G., Zentner, R. P.,Selles, F. and Curtin, D. 1996. Tillage and crop rotation effects on soil organic C and N in a coarse-textured typic haploboroll in Southwestern Saskatchewan. *Soil and Tillage Research* 37, 3–14.

Campbell, J. P. Sr. 2005. *Improving Soil Productivity with Conservation Tillage and Double Cropping: A History of the P1 Watershed*. Watkinsville, GA: Agricultural Research Service, South Atlantic Area Resource Conservation Center.

Chepil, W. S. and Woodruff, N. P. 1963. The physics of wind erosion and its control. *Advances in Agronomy* 15, 211-302.

Conservation Technology Information Center. 2023. *CRM Survey Data Tillage Practices*. CRM: Conservation Technology Information Center (ctic.org).

Dao, T. H. 1993. Tillage and winter wheat residue management effects on water infiltration and storage. *Soil Science Society of America Journal* 57, 1586–1595.

Duniway, M. C., Pfennigwerth, A. A., Fick, S. E., Nauman, T. W., Belnap, J. and Barger, N. N. 2019. Wind erosion and dust from US drylands: a review of causes, consequences and solutions. *Ecosphere* 10(3), e02650.

Evett, S. R., Peters, F. H., Jones, O. R. and Unger, P. W. 1999. Soil hydraulic conductivity and retention curves from tension infiltrometer and laboratory data. In M. T. van Genuchten, F. J. Leij, and L. Wu (Eds.), *Proceedings of the International Workshop on Characterization and Measurement of the Hydraulic Properties of Unsaturated Porous Media*. Riverside: University of California.

Francis, G. S. and Knight, T. L. 1993. Long-term effects of conventional and no-tillage on selected soil properties and crop yields in Canterbury, New Zealand. *Soil and Tillage Research* 26,193-210.

Franzluebbers, A. J. 2002. Water infiltration and soil structure related to organic matter and its stratification with depth. *Soil and Tillage Research* 66, 197–205.

Franzluebbers, A. J. and Stuedemann, J. A. 2008. Soil physical responses to cattle grazing cover crops under conventional and no tillage in the Southern Piedmont, USA. *Soil and Tillage Research* 100, 141–153.

Fryrear, D. W. 1985. Soil cover and wind erosion. *Transactions of the ASAE* 28(3), 781–784.
Gao, L., Wang, B., Li, S., Wu, H., Wu, X., Liang, G., Gong, D., Zhang, X., Cai, D. and Degré, A. 2019. Soil wet aggregate distribution and pore size distribution under different tillage systems after 16 years in the Loess Plateau of China. *Catena* 173, 38–47.
Gollany, H. 2022. Assessing the effects of crop residue retention on soil health. In W. Horwath (Ed.), *Improving Soil Health*. Cambridge, UK: Burleigh Dodds Science Publishing.
Hamm, A. and Manter, D. K. 2022. Assessing the effects of no-till cultivation practices on soil health. In W. Horwath (Ed.), *Improving Soil Health*. Cambridge, UK: Burleigh Dodds Science Publishing.
Harrold, L. L. and Edwards, W. M. 1974. No-tillage system reduces erosion from continuous corn watersheds. *Trans ASAE* 17(3): 0414-0416.
Heike, F. and Schmidt, J. 2013. The effect of soil bulk density on rill erosion. *Zeitschrift für Geomorphology: Supplementary Issue* 57(2), 256–266.
Hubbard,R. K., Strickland, T. C. and Phatak, S. 2013. Effects of cover crop systems on soil physical properties and carbon/nitrogen relationships in the coastal plain of Southeastern USA. *Soil and Tillage Research* 126, 276–283.
Hudson, B. D. 1994. Soil organic matter and available water capacity. *Journal of Soil and Water Conservation* 49(2), 189–194.
Jabro, J., Evans, R., and Iversen, W. 2012. Freeze-Thaw Cycles Effects on Soil Compaction in Clay Loam. *Geophysical Research Abstracts*. Vol. 14, EGU2012-1688.
Jemai, I., Aissa, N. B., Guirat, S. B., Ben-Hammouda, M. and Gallali, T. 2013. Impact of three and seven years of no-tillage on the soil water storage, in the plant root zone, under a dry subhumid Tunisian climate. *Soil and Tillage Research* 126, 26–33.
Jepson, R., Roberts, J. and Lick, W. 1997. Effects of bulk density on sediment erosion rates. *Water, Air and Soil Pollution* 99, 21–31.
Jiang, P., Anderson, S. H., Kitchen, N. R., Sadler, E. J. and Sudduth, K. A. 2007. Landscape and conservation management effects on hydraulic properties of a claypan-soil toposequence. *Soil Science Society of America Journal* 71, 803–811.
Kahlon, M. S., Lal, R. and Varughese, M. A. 2013. Twenty-two years of tillage and mulch impact on physical characteristics and carbon sequestration. *Soil and Tillage Research* 125, 151–158.
Karlen, D. L., Wollenhaupt, N. C., Erbach, D. C., Berry, E. C., Swan, J. B., Eash, N. S. and Jordah, J. L. 1994. Crop residue effects on soil quality following 10 years of no-till corn. *Soil and Tillage Research* 31, 149–167.
Krauss, M., Wiesmeier, M., Don, A., Cuperus, F., Gattinger, A., Gruber, S., Haagsma, W. K., Peigné, J., Palazzoli, M. C., Schulz, F., van der Heijden, M. G. A., Vincent-Caboud, L., Wittwer, R. A., Zikeli, S. and Steffens, M. 2022. Reduced tillage in organic farming affects soi organic carbon stock in temperate Europe. *Soil and Tillage Research* 216, 105262.
Li, Y.X., Tullberg, J. N., Freebairn, D. M., McLaughlin, N. B. and Li, H. W. 2008. Effects of tillage and traffic on crop production in dryland farming systems: I. evaluation of PERFECT soil-crop simulation model. *Soil and Tillage Research* 100, 15–24.
Moebius-Clune, B. A., van Es, H. M., Idowu, O. J.,Schindelbeck, R. R., Moebius-Clune, D. J., Wolfe, D. W., Abawi, G. S., Thies, J.E., Gugino, B. K. and Lucey, R. (deceased). 2008. Long-term effects of harvesting maize stover and tillage on soil quality. *Soil Science Society of America Journal* 72, 960–969.

Nordstrum, K. F. and Hatta, S. 2004. Wind erosion from cropland in the USA: a review of problems, solutions and prospects. *Geoderma* 121(3–4), 157–167.

Olson, K. R., Ebelhar, S. A. and Lan, J. M. 2013. Effects of 24 years of conservation tillage systems on soil organic carbon and soil productivity. *Applied and Environmental Soil Science* 2013, Article ID 617504.

Panagos, P., Borrelli, P., Meusburger, K., Alewell, C., Lugato, E. and Montanarella, L. 2015. Estimating the soil erosion cover-management factor at the European scale. *Land Use Policy* 48(2015), 38–50.

Raczkowski, C. W., Mueller, J. P., Busscher, W. J.,Bell, M. C. and McGraw, M. L. 2012. Soil physical properties of agricultural systems in a large-scale study. *Soil and Tillage Research* 119, 50–59.

Reynolds,W. D., Drury, C. F., Yang, X. M., Fox, C. A., Tan, C. S. and Zhang, T. Q. 2007.Land management effects on the near-surface physical quality of a clay loam soil. *Soil and Tillage Research* 96, 316–330.

Rhoton, F. E., Bruce, R. R., Buehring, N. W., Elkins, G. B., Langdale, C. W. and Tyler, D. D. 1993. Chemical and physical characteristics of four soil types under conventional and no-tillage systems. *Soil and Tillage Research* 28, 51–61.

Rickson, J. 2018. Mechanisms of soil erosion/degradation. In D. Reicosky (Ed.), *Managing Soil Health for Sustainable Agriculture Volume 1: Fundamentals*. Cambridge, UK: Burleigh Dodds Science Publishing.

Sasal, M. C., Castiglioni, M. G. and Wilson, M. G. 2010. Effect of crop sequences on soil properties and runoff on natural-rainfall erosion plots under no tillage. *Soil and Tillage Research* 108, 24–29.

Schwen, A., Bodner, G., Scholl, P., Buchan, G. D.and Loiskandl, W. 2011. Temporal dynamics of soil hydraulic properties and the water-conducting porosity under different tillage. *Soil and Tillage Research* 113, 89–98.

Sharratt, B., Zhang, M. and Sparrow, S. 2006. Twenty years of conservation tillage research in Subarctic Alaska. II. Impact on soil hydraulic properties. *Soil and Tillage Research* 91, 82–88.

Smith, C. and Kucera, M. 2018. *Effects on Soil Water Holding Capacity and Soil Water Retention Resulting from Soil Health Management Practices Implementation*. USDA NRCS Website AWC Effects on Soil Water Holding Capacity and Soil Water Retention Resulting from Soil Health Management Practices (usda.gov)

So, H. B., Grabski, A. and Desborough, P. 2009. The impact of 14 years of conventional and no-till cultivation on the physical properties and crop yields of a loam soil at Grafton NSW, Australia. *Soil and Tillage Research* 104, 180–184.

Stipesevic, B. and Kladivko, E. J. 2002. *The Effects of Winter Wheat Cover Crop Desiccation Times on Soil Physical Properties and Early Maize Growth in No-Till and Conventional Tillage*. Trojstva, Osijek, Croatia: University of J.J. Strossmayer, Faculty of Agronomy.

Strudley, M. W., Green, T. R. and Ascough II, J.C.2008. Tillage effects on soil hydraulic properties in space and time: state of the science. *Soil and Tillage Research*99, 4–48.

Stubbs, T. L., Kennedy, A. C. and Schillinger, W.F. 2004. Soil ecosystem changes during the transition to no-till cropping. *Journal of Crop Improvement* 11, 105–135.

Tatarko, J., Wagner, L. and Fox, F. 2019. *Wind Erosion Processes and Control*. Colorado State University Extension Publication. https://www.ars.usda.gov/research/publications/publication/?seqNo115=342843

USDA. 2013. *Science Documentation: Revised Universal Soil Loss Equation Version 2*. Washington, DC: USDA-ARS.

USDA. 2023. *Wind Erosion Prediction System (WEPS) Version 1.5*. Wind Erosion Prediction System (usda.gov) website.

USDA. 2024. *USDA-WEPP Hillslope Profile and Watershed Model Documentation NSERL Report #10*. WEPP Model Documentation: USDA ARS website.

USDA-NRCS National Handbook of Conservation Practices. 2024. Conservation Practice Standards | Natural Resources Conservation Service (usda.gov)

USDA-NRCS. 2015. *Soil Health Literature Summary-Effects of Conservation Practices on Soil Properties in Areas of Cropland*. USDA National Soil Survey Center (lead author Michael J. Kucera Agronomist USDA-NRCS). https://www.nrcs.usda.gov/sites/default/files/2022-10/Soil_Health_Literature_Summary_Effects_of_Conservation_Practices_Jan2015.pdf

Van Doren, D. M. and Triplett, G. B. 1982. *Crop Production Demands and the Conservation of Soil Resources in Ohio*. Columbus, OH: Ohio State University publications.

Van Doren, D. M., Moldenhauer, W. C. and Triplett, G. B. 1984. Influence of long-term tillage and crop rotation on water erosion. *Soil Science Society of America Journal* 48(3), 636–640.

Villamil, M. B., Bollero, G. A., Darmody, R. G., Simmons, F. W. and Bullock, D. G. 2006. No-till corn/soybean systems including winter cover crops: effects on soil properties. *Soil Science Society of America Journal* 70, 1936–1944.

Wells, R., Flanagan, D. C., Langendoen, E. J., McGehee, R. P., Bingner, R. L., Frankenberger, J. R., Locke, M. A., Momm, H. G., Renschler, C. S., Srivastava, A., Vieira, D. and Tsegaye, T. 2024. Cropland erosion estimates simulated by RUSLE2 and WEPP: results from two initial studies. *Journal of Soil and Water Conservation* (In Review).

Wilson, G. V., McGregor, K. C. and Boykin, D. 2008. Residue impacts on runoff and soil erosion for different corn plant populations. *Soil and Tillage Research* 99(2), 300–307.

Wischmeier, W. H. and Smith, D. D. 1965. *Predicting Rainfall-erosion Losses from Cropland East of the Rocky Mountains*. Agriculture Handbook NO. 282. USDA-ARS in Cooperation with Purdue Agricultural Experiment Station.

Wischmeier, W. H. and Smith, D. D. 1978. *Predicting Rainfall Erosion Losses, a Guide to Conservation Planning*. Supersedes Agriculture Handbook NO. 282. USDA-ARS in Cooperation with Purdue Agricultural Experiment Station.

Zhang, G. S., Chan, K.Y., Li, G. D. and Huang, G. B. 2008. Effect of straw and plastic film management under contrasting tillage practices on the physical properties of an erodible loess soil. *Soil and Tillage Research* 98, 113–119.

Chapter 9

The use of soil stabilizers to prevent erosion

G. J. Levy, Agricultural Research Organization, Israel; and A. I. Mamedov, Tottori University, Japan

1 Introduction

Soil erosion, especially by water but also by wind, is a principal cause of soil degradation and the delivery capacity of its ecosystem services. Consequently, soil erosion is a major global threat to the ability of agriculture to ensure human needs, including food, feed, fiber, clean water and other soil ecosystem services in the twenty-first century.

Mitigation of soil erosion is, therefore, essential for sustaining the aforementioned services provided by soil. Common strategies for soil erosion mitigation include (e.g. Graber et al., 2006 and references cited therein):

- Traditional or Conservation Agriculture techniques such as minimum tillage;
- Use of residue material and/or cover crop to protect the soil surface from raindrop impact;
- Enhancement of effective soil surface roughness to reduce runoff development, volume and velocity; and
- Modification of slope-length gradient and direction of runoff flow.

These conservation approaches and others are discussed in detail in other chapters of this book.

http://dx.doi.org/10.19103/AS.2023.0131.18

An alternative approach for mitigating soil erosion, particularly on cultivated land, is one that advocates modification of soil properties associated with processes that influence soil structure stability degradation (Levy and Warrington, 2015 and references cited therein). By enhancing the stability of surface aggregates and preventing soil clay dispersion, it is possible to maintain soil permeability to water and reduce soil susceptibility to crust formation, thereby decreasing runoff soil erosion. Judicious application of soil-stabilizing agents can contribute to enhanced structural stability of the soil surface.

This chapter discusses the potential of three types of soil-stabilizing amendments, namely gypsum and gypsiferous materials (GM), synthetic polymers and organic matter waste materials. The review focuses on the impact of these amendments on cultivated soils, particularly those with poor structural stability (e.g. soils in semiarid or humid regions) that are characterized by low organic matter content and/or elevated sodicity levels, which often result in limited infiltration, runoff initiation and soil erosion (NRCS, 2015; Levy and Nachshon, 2022).

2 Gypsum and gypsiferous materials as soil stabilizers

Gypsum ($CaSO_4 \bullet 2H_2O$), from both natural and anthropogenic sources (byproducts), is a commonly used amendment for improving soil structure stability, especially in poorly structured soils (e.g. sodic soils), as well as an ameliorant to counter subsoil acidity and as a nutrient source for Ca. This section focuses on gypsum as a soil-stabilizing agent. Gypsiferous material (GM) from industrial sources has a higher dissolution rate compared with mined gypsum (Keren and Shainberg, 1981). It is also cheaper and plentiful and therefore preferable as a soil amendment (Levy, 2012). Typically, GM includes phosphogypsum (PG) obtained from the phosphorus industry (Shainberg et al., 1989), as well as flue gas desulfurization gypsum (FDG) and fluidized bed combustion bottom ash (FBCBA) obtained from desulfurization of stack gases from coal-fired power-producing facilities (Norton et al., 1993).

When GM is added to poorly structured soils, it can increase aggregate stability and prevent clay dispersion and swelling (Lebron et al., 2002; Mamedov et al., 2007) by means of increasing the soil solution electrolyte concentration (EC) and cation exchange effects (Loveday, 1976). The relative importance of each of these effects depends on the purpose for which GM is used. To control soil surface permeability, runoff and erosion, GM is used to stabilize surface aggregates during soil exposure to achieve suitable water quality (i.e. rainwater or another source of water with low EC), making the electrolyte effect of greater importance. In this case, surface application of GM is recommended, and the dissolution rate of the GM dictates whether the rate of electrolyte supply to the water is sufficient to prevent clay dispersion and swelling. In this case, the amount of GM required depends on the annual amount of high-quality water applied and the rate of GM dissolution. Conversely, in soils where the soil profile needs to be remediated, e.g. in cases of sodic soils, other considerations are important. The amount of GM required may depend on the initial and desired final exchangeable Na levels, soil cation exchange capacity, bulk density and depth of soil to be reclaimed (Levy and Nachshon, 2022).

Numerous studies have reported that GM addition to the soil surface can significantly decrease soil loss (e.g. Graber et al., 2006, and references cited therein). Laboratory studies have indicated that adding GM to the soil surface, commonly at the rate of 5 Mg ha^{-1}, decreased soil loss by 30–60% compared with non-treated soils (e.g. Flanagan et al., 1997; Peterson et al., 2002). Under field conditions, GM efficiency in controlling soil erosion was of a similar magnitude to that observed in laboratory studies in highly weathered soils from the coastal plain in southeastern US, characterized by a humid subtropical climate (Truman et al., 2010), but in semiarid soils, the use of GM had a much greater impact, resulting in three to five times lower erosion levels than those from the control plots (e.g. Agassi et al., 1985).

Transitioning from gentle-to-steep slopes increases the dominance of rill-type erosion over interill erosion (Kimaro et al., 2008). In the latter, erosion is controlled by concentrated flow in the rills (Govers et al., 2007). For soils susceptible to sealing, slope has a complex effect on runoff, infiltration rate and erosion (Poesen, 1987). On steep slopes, runoff velocity and erosion increase. However, since the seal itself may be eroded, it is possible that with an increase in slope and seal erosion, the infiltration rate may increase and runoff decrease. Warrington et al. (1989) investigated the effect of GM on erosion of a sandy loam soil on seven slopes with gradients ranging between 5% and 30% using a laboratory rainfall simulator (Morin et al., 1967). Soil samples were packed in 10 cm deep, 30 × 50 cm metal boxes, over the surface of which GM was spread at an equivalent rate of 5 Mg ha^{-1}. The soil samples were then subjected to 90 mm of simulated rain comprising deionized water. The final infiltration rate increased for both treated and untreated soil with an increase in slope (Fig. 1a). Intensifying the slope from 5% to 25% increased soil loss by just 190% for phosphogypsum (a type of GM)-treated soil compared to a 670% increase in erosion in the untreated soil (Fig. 1b). It should be noted that (1) erosion in the GM treatment increased linearly with the increase in slope and (2) the favorable effect of GM on reducing soil erosion, compared to erosion obtained in the control, occurred on slopes >10% (Fig. 1b). Warrington et al. (1989) observed the formation of rills on the untreated soil at slopes >10% which were accompanied by a substantial increase in soil loss. On slopes >10%, rill erosion was the dominant process for erosion in the non-treated soil, while under the GM treatment, interill erosion was maintained for all slopes. The results of laboratory studies by

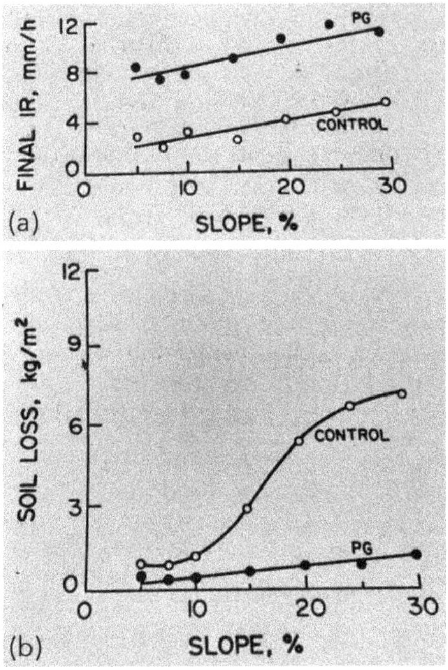

Figure 1 Effects of slope and phosphogypsum (PG) on (a) final infiltration rate and (b) soil loss. Source: Adapted from: Warrington et al. (1989).

Warrington et al. (1989) have been validated in field studies (Agassi et al., 1990). The efficiency of GM in reducing soil loss on steep slopes was found to depend on slope length, with a much greater effect on shorter slopes (usually tested under laboratory conditions) in which interrill erosion, rather than rill erosion, was the predominant process, even when GM was used (Agassi and Ben-Hur, 1991).

Based on these results, it can be concluded that surface spreading of GM releases electrolytes to the percolating and runoff water and thus decreases soil erosion by the following mechanisms:

1 The amount of rain that penetrates the soil increases, so that runoff depth decreases.
2 The stability of surface aggregates increases and fewer particles are detached by either raindrop impact or overland flow.
3 The velocity of runoff flow decreases by maintaining a rough surface at the soil surface; and
4 The flocculation of suspended clay size particles and the resultant deposition is enhanced.

In addition, Agassi et al. (1986) suggested that the favorable effect of surface-spread GM at an equivalent rate of 5 Mg ha^{-1} should also be attributed to (1) physical interference with the continuity of the seal and (2) partial mulching of the surface, protecting it from the beating action of raindrops.

3 Polymers as soil stabilizers

The use of synthetic organic polymers as a soil conditioner began in the 1950s. Numerous studies have discussed the use of organic polymers as soil conditioners in general (e.g. Azzam, 1980) and for improving soil structure and physical properties in particular (Wallace, 1998; Sojka et al., 2007; Levy and Warrington, 2015). Of the various polymers studied over the years, water-soluble (i.e. non-crosslinked) polyacrylamide (PAM) has received considerable attention due to its impact on various issues associated with soil structure and stability. Due to fears that PAM may contain residual amounts of, or degrade to produce, acrylamide, which is a nerve agent, only food-grade anionic PAM is recommended for use in agricultural soils (Sojka et al., 2007). This type of PAM has very little residual acrylamide (<0.05%). Furthermore, the relatively small quantities used present minimal risk of contamination to the environment as anionic PAM is generally not desorbed from the soil and is not likely to enter the water system (Nadler et al., 1992). Studies have also shown that PAM itself does not pose a threat to human health (Seybold, 1994) and that even when subjected to thermal degradation while in liquid form does not release harmful acrylamide in levels that would cause drinking water to exceed the US Environmental Protection Agency standards of 0.5 ppb (Woodrow et al., 2008). Therefore, anionic PAM soil stabilizers are considered to be safe and environmentally friendly.

Over the last three decades, anionic PAM, of generally high molecular weight (~20×10^6 Da) with 10–30% charge density, has been studied and used intensively in commercial fields for controlling irrigation-induced erosion and enhancing infiltration (e.g. Mamedov et al., 2009; Levy and Warrington, 2015). The main benefit of the addition of PAM to soil surface aggregates lies in the stabilization of existing aggregates and improved bonding between adjacent aggregates, mainly through adsorption on the outer surfaces of soil particles (Fig. 2).

Use of PAM to control runoff and soil erosion has been studied extensively under both rain and irrigation (furrow and overhead) conditions. PAM can be applied via the irrigation water, by spraying a PAM solution onto the soil surface, or as a dry powder spread on the soil surface and subsequently dissolved by rain or irrigation water (Levy and Warrington, 2015). In the case of rain or overhead irrigation, addition of PAM to the soil mitigates the adverse effects of the beating action of the water drops on the surface aggregates. Conversely, in

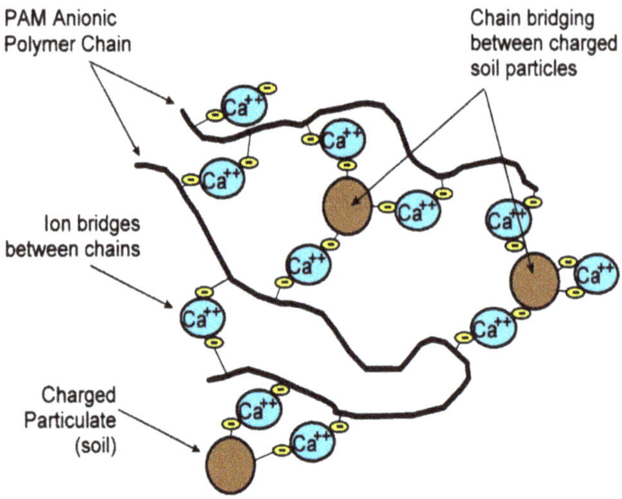

Figure 2 Schematic description of interactions of anionic PAM with soil particles. Source: Adapted from: Orts et al. (2002).

furrow irrigation, PAM is primarily added to enhance the stability of the surface aggregates against the shearing force of the flowing water, thereby preventing furrow erosion.

3.1 Furrow irrigation

Erosion caused by surface/furrow irrigation is a severe problem affecting millions of hectares of irrigated cropland (Sojka and Surapaneni, 2000). Laboratory and field tests, mostly in the US, clearly indicate that the use of PAM in furrow irrigation leads to an almost total prevention of erosion and to a significant increase in water infiltration (Lentz and Sojka, 2000). The beneficial effects of PAM application in furrow irrigation, coupled with its relative ease of application, have resulted in rapid acceptance of the technology in the US. At the end of the twentieth century, over 400 000 ha of irrigated land employed PAM for erosion and/or infiltration management (Sojka and Surapaneni, 2000).

The rapid acceptance of this technology can be attributed to several factors. The amount of PAM required to control erosion in furrow irrigation is <1.0 kg ha^{-1} (Lentz and Sojka, 1994), with effective concentrations of PAM in irrigation water being 2–10 mg L^{-1}. These amounts are significantly smaller than the amounts required to control sealing and erosion in dryland farming.

For PAM application to be effective in controlling furrow erosion, it is imperative to add it with the first irrigation on a newly tilled field. Following initial PAM treatment, erosion in subsequent irrigations can generally be controlled with much less than 10 mg L^{-1} (1-5 mg L^{-1} PAM) if the soil has not been disturbed between irrigations. Application of PAM to surfaces containing aggregates that have been slaked (e.g. by raindrops) minimizes PAM effectiveness (Wallace, 1998).

3.2 Rain and overhead irrigation conditions

Research into the efficacy of soil surface application of polymers (either in the form of dry grains or by spraying a concentrated PAM solution) has shown that an addition of 10-40 kg PAM ha^{-1} is required to maintain significantly lower levels of runoff and erosion in soils exposed to rain as compared with non-treated soils. For the control of runoff and erosion in sprinkler irrigation, the amounts required are lower, at less than 5 kg PAM ha^{-1} (Levy and Warrington, 2015). The difference in the required amount of PAM may stem from the fact that the kinetic energy of raindrops is usually higher than that of water drops in sprinkler irrigation.

It is important to note that, under rain conditions, the addition of PAM to the soil surface in the form of dry grains is effective in controlling soil erosion but ineffective in controlling runoff (Fig. 3). Yu et al. (2003) suggested that, upon dissolution of PAM grains by rainwater, PAM molecules adsorb on the surface aggregates, binding them together to form a cohesive surface that better

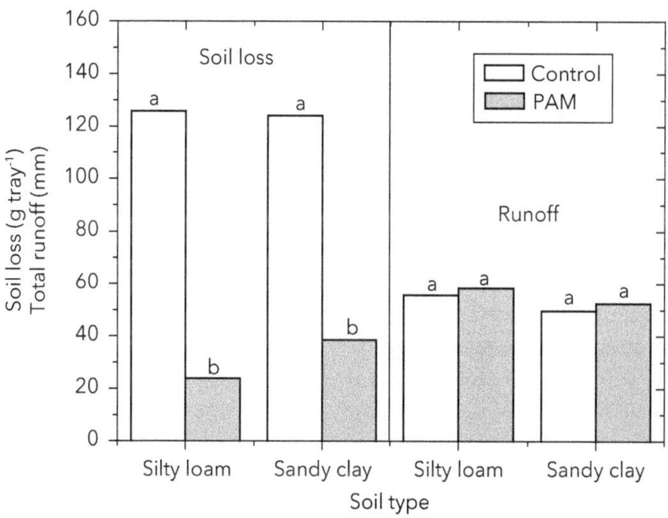

Figure 3 Soil loss and total runoff for untreated (control) or PAM-treated (20 kg ha^{-1}) soil samples of a silty loam and a sandy clay. Within a soil type and type of measurement, bars labeled with the same letter do not differ significantly at the $P < 0.05$ level. Source: Modified from Yu et al. (2003).

resists particle detachment by the raindrops and thus reduces soil erosion. At the same time, PAM molecule segments that do not adsorb on the surface particles instead block the water conducting pores, resulting in the observed trend of a lower infiltration rate, which leads to higher runoff levels.

Studies in which a surface application of PAM (irrespective of the method of application) was supplemented with gypsum showed a significantly lower runoff and erosion than those in which either no amendments (control) or an application of just one of the amendments were used (Levy and Warrington, 2015, and references cited therein). In the presence of electrolytes (e.g. gypsum solution), the negative charge and the thickness of the diffuse double layer at the clay and polymer surfaces are suppressed, resulting in decreased repulsion forces and greater adsorption of soil particles to the anionic polymer. In addition, the dissolved gypsum increases the EC in the soil solution to values above the flocculation value of the soil clay (Smith et al., 1990). Moreover, the application of PAM in combination with gypsum is highly effective in reducing runoff and erosion in highly sodic smectitic soils, i.e. highly dispersible soils (Tang et al., 2006; Mamedov et al., 2020), as well as in a kaolinitic acidic soil (Lee et al., 2010). The effectiveness of treating the soil with a combination of PAM and gypsum to decrease erosion is related to the reduction of both the physical disintegration of surface aggregates by PAM and the chemical dispersion of the soil clay by gypsum.

Conflicting results have been reported on the residual effects of PAM addition. Some studies have reported that addition of PAM via irrigation water at rates of 10–30 mg L^{-1} is effective at maintaining high infiltration rates and low levels of soil erosion in subsequent irrigations (e.g. Aase et al., 1998). Other researchers observed that the addition of low concentrations (10–20 mg L^{-1}) of PAM in the irrigation water had no effect on soil erosion in subsequent irrigations with water alone (Levy et al., 1992; Gardiner and Sun, 2002).

3.3 Structure stability

The impact of PAM on soil erosion can also be inferred from its effects on soil structure stability. The existence of a significant inverse linear relation ($R^2 = 0.81$) between a structure stability index (a proxy for aggregate stability) and soil erosion has been reported (Norton et al., 2006; Mamedov et al., 2021a). Such studies reveal that PAM addition to a conventionally tilled soil can improve its stability by two to three times that of a stable non-tilled soil (Mamedov et al., 2021a). PAM stabilizing effects depend on soil inherent properties and are more pronounced in *a priori* poorly structured soils, e.g. soils with low organic matter content (Caesar- Tonthat et al., 2008; Mamedov et al., 2021b) and soils with coarse texture (Levy and Mamedov, 2013; Mamedov et al., 2020). In addition, PAM efficacy in stabilizing soil aggregates depends on soil clay mineralogy in the following increasing order: smectite > illite > kaolinite (Mamedov et al., 2010). However, Mamedov et al. (2017) warned that the stabilizing effects of PAM depend on conditions prevailing in the field (e.g. size of the soil surface aggregates) and hence PAM should not be considered a general soil-stabilizing agent.

4 Organic residues as soil stabilizers

4.1 Role of organic matter in soil stabilization

The contribution of soil organic matter (SOM) to soil structural stability and decreased soil erosion is often characterized by the effects of SOM on aggregate stability. The role of organic matter in the formation and maintenance of water stable soil aggregates has been described in research by Tisdall and Oades (1982). In their study, organic binding agents are classified into three groups as a function of longevity of the binding agent and its effect on aggregate stability. As a result, a common approach to improving soil structural form and stability, especially in semiarid and arid regions where SOM content is *a priori* low, is the incorporation of organic residues into cultivated soils (e.g. Angers and Carter, 1996). Archaeological evidence shows this practice goes back more than 2500 years (Parr and Hornick, 1992). The addition of organic residues enhances soil microbial biomass activity, which transforms newly added organic matter into materials such as polysaccharides and long chain aliphatic compounds that act to bind and stabilize aggregates (Perucci, 1990; Plante and Voroney, 1998). Microorganisms also promote aggregation through the binding properties of their bodies, such as fungal filaments (Tisdall and Oades, 1982) or bacterial gums and exudates. A comprehensive review of the role of organic matter in promoting soil structure can be found in Carter and Stewart (1995). However, as already stated by Kay and Angers (2000) and highlighted in the current review, exceptions to the commonly accepted notion that increased SOM content (via the addition of various sources and types of organic wastes) tends to improve soil structure stability may occur. This is because this newly modified SOM is not a homogeneous substance; it is comprised of plant, animal and microbial materials in various stages of decomposition and consequently associated in a variety of ways with the amount of soil minerals.

4.2 Animal wastes

Animal wastes are a major global source of organic residues, which can be applied to cultivated soils (e.g. Martinez et al., 2009). Animal wastes are produced in large quantities and can pose a threat to the environment if not disposed of appropriately. Every year, around 35 million tons of dry livestock waste is produced in the US alone, and in Europe more than 1500 million tons of fresh livestock waste is generated annually (Ilea, 2009; Zhu and Hiltunen, 2016). Use of animal wastes as a source of amending cultivated land is therefore essential, as it can serve a dual purpose of disposing of the waste in a sustainable way and potentially acting as a soil amendment agent.

Many studies have reported the advantageous reduction effects of soil erosion via enhancing soils with animal waste (e.g. Chandra and De, 1982; Gilley and Risse, 2020; Tejada and Gonzalez, 2007; Nasir Ahmed et al., 2020 and references cited therein). For instance, Gilley and Risse (2020) in the US reported a decrease of 15–65% in soil erosion (under natural rain conditions) in seven different manure-amended soils compared with non-amended ones. The favorable impact of animal wastes was ascribed to (1) its high organic matter content that made the soil more resistant to detachment and (2) the improvement of soil structure via enhancement of soil aggregation.

There are, however, notable exceptions in the literature in which contradictory results to the aforementioned findings have been reported. Whalen and Chang (2002) found hardly any effect on the mass of aggregates smaller than 2 mm, while for larger macroaggregates (>2 mm), they found that fresh manure tended to reduce macroaggregate mass. The increased breakdown of these macroaggregates was attributed to dispersion processes, due to the appreciable content of monovalent cations (Na^+ and K^+) in the manure, and/or to other foreign dispersing agents in the manure. Based on their findings, Whalen and Chang (2002) suggested that long-term fresh manure application may result in greater risk to soil loss to erosion, particularly for dryland agriculture.

A recent laboratory study by Goldberg et al. (2020) also reported that addition of cattle manure increased soil erosion compared to no manure addition, with the extent of this phenomenon dependent on soil type (Fig. 4).

The negative effects of manure on soil erosion are explained by the fact that the erosion measurements using a laboratory rainfall simulator (Morin et al., 1967) took place just 21 days after manure addition, and thus the decomposing processes of the applied organic residues and assimilation of the products into the soil aggregates was insufficient to stabilize the latter (Goldeberg et al., 2020).

The addition of composted wastes to cultivated lands can serve as a viable alternative to the addition of raw organic wastes that could at times have environmentally harmful consequences. Compost consists of relatively stable decomposed organic materials resulting from the accelerated biological degradation of organic materials under controlled, aerobic conditions (Paulin and O'Malley, 2008). Although there seems to be a general consensus (e.g. Aggelides and Londra, 2000) that mature compost has a lesser effect on enhancing soil structure stability in comparison with readily decomposable organic matter, numerous studies have reported significant positive effects for

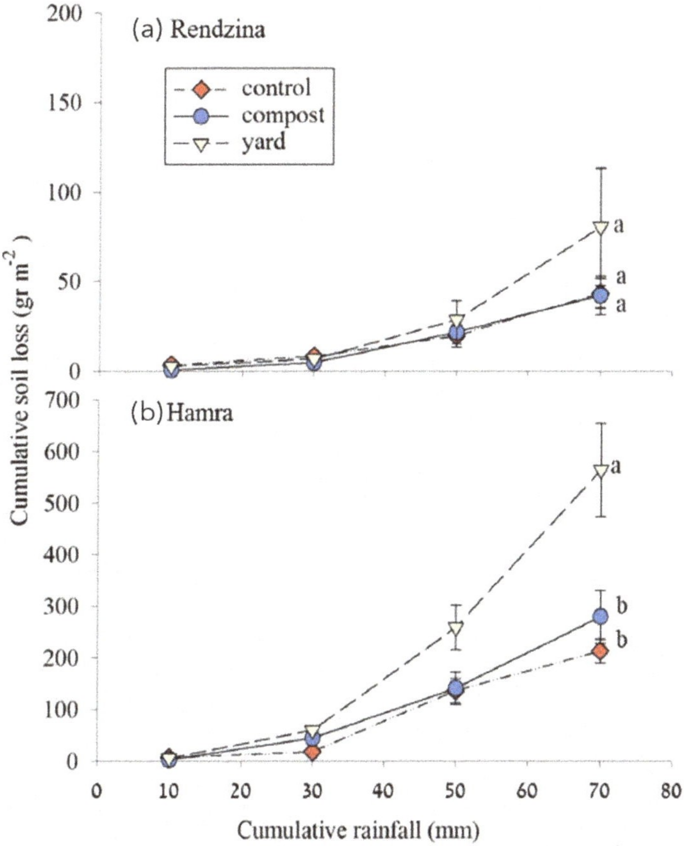

Figure 4 Cumulative soil loss for Rendzic Leptosols (Rendzina) (a) and Vertic Luvisol (Hamra) (b) soils, with no addition of organic matter (control), and addition of yard manure (yard) and compost. Different letters 'a' and 'b' indicate significant differences between treatments. Source: Adapted from: Goldberg et al. (2020).

compost application to soils, especially those with initially low SOM content (e.g. reviews by Hargreaves et al., 2008 and Adugna, 2016 and references cited therein). A study by Bhattarai et al. (2011) compared the impact of compost addition on soil erosion under both laboratory (at slopes of 4, 8 and 16%) and field (3.5% slope) conditions and observed that compost addition decreased erosion by >50% compared to bare soil. Evidently, compost quality, including the quality of the original raw material, its dose, the maturation degree, C/N ratio, together with conditions prevailing in the field (e.g. soil type and slope) all determine the magnitude of the efficacy of the addition of compost on decreasing soil erosion (Martínez-Blanco et al., 2013). However, there are also reports in which a single application of compost did not significantly influence runoff and erosion under simulated rainfall conditions (Gilley and Eghball, 1998).

4.3 Biosolids

Biosolids are sewage sludges (i.e. the solid mass obtained in the course of wastewater treatment) that have undergone treatment in accordance with the necessary regulations to permit its application to land. To meet the regulatory requirements for land application, sewage sludges must undergo treatments including aerobic digestion, anaerobic digestion, composting, alkaline stabilization and thermal drying (Qin et al., 2012).

A few decades ago, biosolids were more frequently used for reclamation of highly eroded soils or strip-mined lands than cultivated agricultural land (Graber et al., 2006). Compared with examining biosolid use for establishing, maintaining and improving plant growth and crop yield and determining accumulation and leaching of toxic elements, salts and other xenobiotics, fewer studies of land-applied biosolids on cultivated land have reported the impact of biosolids on soil structure stability (Qin et al., 2012). Illera et al. (1999) examined the effects of anaerobically digested biosolids that were surface-applied to a degraded highly carbonated soil in Spain on the physical and chemical properties of soil. Following 12 months of biosolid application, the structure of soil was improved to a depth of 30 cm due to decreased bulk density and increased water retention. Similarly, results from a field trial in a semiarid region of Greece revealed that the addition of biosolids, added to the soil surface and rototilled into the top 15 cm, enhanced aggregate stability, and thus soil stability in two soil types, by around 30% for loam soil and a more modest 17% for clay, at the highest application rate of 300 m^3 ha^{-1} (Aggelides and Londra, 2000). Similar observations were reported by Moffet et al. (2005) on soils in the US and Garcia-Orenes et al. (2005) on two soils in Spain. In their review, Qin et al. (2012, and references cited therein) concluded that the soil's improved physical properties and structure stability following the application of biosolids coincided with an increase in soil organic C.

4.4 Biochar

Biochar is a fine-grained, porous, black carbonaceous product resulting from slow and incomplete combustion of various sources of biomass, including agroforestry, animal wastes, animal manures, sewage sludge and invasive plant species (Wani et al., 2021). The production of biochar involves pyrolyzing biomass under controlled temperature and pressure conditions. The characteristics of biochar depend upon the pyrolysis type, biomass type and preparation and the conditions adopted during the process (Ramola et al., 2020). Biochar contains around 60–80% carbon and is, therefore, considered a good way of enhancing soil properties (Blanco-Canqui, 2017). Overall, biochar improves the soil's physical status through decreasing soil bulk density by 3-31%, increasing porosity by 14-64% and wet aggregate stability by 3-226%. Its impact on dry aggregate stability is inconclusive. The impact of biochar on water flow in soil depends on soil texture, reducing saturated hydraulic conductivity by 7-2270% in coarse-textured soils and increasing it by 25-328% in fine textured soils (Blanco-Canqui, 2017).

Although biochar has generally positive effects on the physical and hydraulic properties associated with the structural stability of soil, contradictory results have been reported in the literature concerning its impact on soil erosion (Brtnicky et al., 2021; Wani et al., 2021). The influence of biochar on reducing soil erosion has been studied across a wide range of biochar application rates (0.1–10% and/or up to 30 Mg ha^{-1}) under both laboratory (e.g. Gholami et al., 2019; Cai et al., 2020; Wei et al., 2023) and field conditions (Lee et al., 2015; Bashagaluke et al., 2019), different soil types (e.g. non-calcareous, calcareous and acidic) and slopes, and varying rain conditions (Wani et al., 2021 and references cited therein). The beneficial effects of biochar on soil erosion have been ascribed to its favorable effects on reducing runoff, increasing vegetation/crop growth covering the soil surface, improving soil surface aggregate stability, facilitating high infiltration rates and low runoff levels and enhancing organic matter content.

The addition of biochar in mostly small doses of up to 4% or 5 Mg ha^{-1} (e.g. Peng et al., 2016; Lee et al., 2018) or with biochar coarse grains >0.5 mm (Wei et al., 2023) yields higher erosion levels than non-amended soils (e.g. Peng et al., 2016; Lee et al., 2018). In one case, it has been postulated that the biochar used (wood-biochar) displayed a hydrophobic nature that led to increased runoff and subsequently more soil erosion (Lee et al., 2018). A similar reasoning (i.e. the development of hydrophobic conditions) was given by Zhang et al. (2019) for increased erosion in biochar-treated sloping croplands of a loess soil. In an additional study on another loess soil, however, the elevated erosion levels upon addition of biochar were ascribed to the soil's inherent poor structural stability (Sadeghi et al., 2020).

The inconclusive results pertaining to the effects of biochar on soil erosion likely stem from the complex interplay of various attributes and how they interact when determining the influence of biochar on soil susceptibility to erosion. Figure 5 shows the different factors that can affect soil sensitivity to erosion, separately as well as combined with other factors. The factors presented in Fig. 5 not only relate to biochar properties (e.g. particle size, porosity and hydrophobicity), the amount added and production method (type of pyrolysis and origin of biomass), but also to soil properties, climatic conditions, biochar placement (soil surface or mixing with top soil layer) and the addition of other amendments such as animal manure and/or fertilizers.

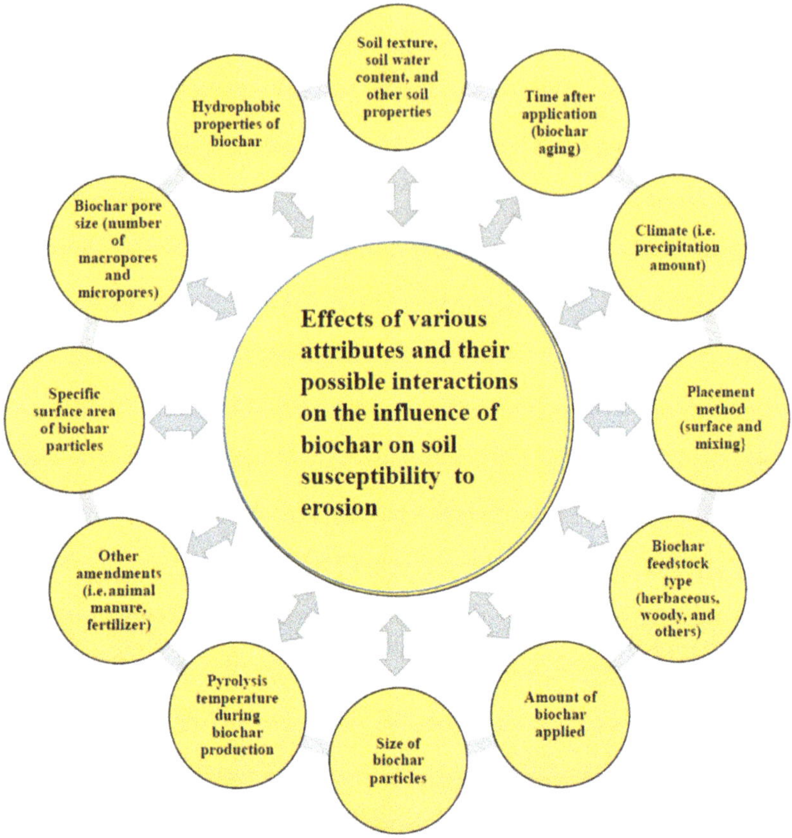

Figure 5 Effects of soil properties, biochar properties and management decisions and their interactions on the influence of biochar on soil susceptibility to erosion. Source: Modified from: Blanco-Canqui (2017).

Knowledge gaps remain in terms of the mechanisms by which biochar alters soil processes. These uncertainties highlight the need for further in-depth and systematic studies, especially under field conditions, on the impact of biochar addition on soil erosion.

4.5 Food-industry wastes

Considerable organic matter-rich wastes are generated in food manufacturing and other industrial processes. In terms of promoting soil structure stability and decreasing soil erosion, the use of unconventional organic waste materials for the joint purpose of waste disposal and soil improvement, including stabilization against erosion, has received considerable attention. A detailed review by Graber et al. (2006) studied the beneficial impacts of food and other industry waste produces, including oily wastes, cheese whey, food-derived polymers, paper industry waste and pulp fibers on soil. The study concluded that the various waste products may contribute to soil stabilization via an increase in amounts of transient, temporary or persistent organic matter.

Depending on the physical and chemical properties of the added amendment and its capacity for decomposition, its mechanism of operation and efficacy may differ substantially.

5 Conclusion

The current review demonstrates that the addition of various amendments to soil can serve as a viable alternative to traditional strategies for enhancing soil structure stability and consequently combating soil erosion. The reviewed amendments enhance soil structure stability by their ability to modify soil properties in such a way that supports the soil's ability to maintain a stable structure. The processes involved include modification of (1) chemical soil properties associated with soil clay stabilization (e.g. use of gypsum and biochar); (2) physical stabilization of soil aggregates (e.g. use of polymers and biochar); and (3) enrichment of SOM (addition of organic waste materials). However, it should be borne in mind that the use of amendments in cultivated lands for alleviating soil erosion requires careful tailoring of the amendment used (including dose, frequency and mode of application) to the soil in question so that the sustainability of the ecosystem and productivity of the treated soil are not compromised.

6 Where to look for further information

Additional sources for information on combating soil erosion in general and by soil stabilizers can be found in the following links:

- How can you help prevent and control soil erosion (https://www.fao.org/family-farming/detail/en/c/1472808/)
- Save our soils: Findings ways to stop erosion (https://www.fao.org/global-soil-partnership/resources/highlights/detail/en/c/1194373/)
- Evaluating the "Soil stabilisation and control of erosion" ecosystem service provided by agricultural ecosystems over the French territory (https://hal.inrae.fr/hal-02738087/file/2019_FAO_GSSE_proceedings_italy_1.pdf)
- *Soil erosion, conservation and rehabilitation* (1995) Agassi, M. (ed), Marcel Dekker NY.

- *Handbook of soil conditioners* (1998). Wallace, A. and Terry, E.R. (eds), Marcel Dekker NY.

7 References

Aase, J. K., Bjornberg, D. L. and Sojka, R. E. (1998). Sprinkler irrigation runoff and erosion control with polyacrylamide–Laboratory tests, *Soil Sci. Soc. Am. J.* 62(6): 1681-1687.

Adugna, G. (2016). A review on impact of compost on soil properties, water use and crop productivity, *Academic Res. J. Agric., Sci. Res.* 4: 93-104.

Agassi, M. and Ben-Hur, M. (1991). Effect of slope length, aspect and phosphogypsum on runoff and erosion from steep slopes, *Austral. J. Soil Res.* 29: 197-203.

Agassi, M., Morin, J. and Shainberg, I. (1985). Effect of raindrop im-pact energy and water salinity on infiltration rates of sodic soils, *Soil Sci. Soc. Am. J.* 49(1): 186-190.

Agassi, M., Shainberg, I. and Morin, J. (1986). Effect of powdered phosphogypsum on the infiltration rate of sodic soils, *Irrig. Sci.* 7(1): 53-61.

Agassi, M., Morin, J. and Shainberg, I. (1990). Slope, aspect and phosphogypsum effects on runoff and erosion, *Soil Sci. Soc. Am. J.* 54(4): 1102-1106.

Aggelides, S. M. and Londra, P. A. (2000). Effects of compost produced from town wastes and sewage sludge on the physical properties of a loamy and a clay soil, *Bioresour. Technol.* 71(3): 253-259.

Angers, D. A. and Carter, M. R. (1996). Aggregation and organic matter storage in cool, humid agricultural soils. In Carter, M. R. and Stewart, B. A. (Eds.), *Structure and Organic Matter Storage in Agricultural Soils*. Boca Raton, FL: Lewis Publishers, 193-211.

Azzam, R. A. I. (1980). Agricultural polymers, polyacrylamide preparation, application and prospects in soil conditioning, *Commun. Soil Sci. Plant Anal.* 11(8): 767-834.

Bashagaluke, J. B., Logah, V., Opoku, A., Tuffour, H. O., Sarkodie-Addo, J. and Quansah, Ch. (2019). Soil loss and run-off characteristics under different soil amendments and cropping systems in the semi-deciduous forest zone of Ghana, *Soil Use Manag.* 35(4): 617-629.

Bhattarai, R., Kalita, P. K., Yatsu, S., Howard, H. R. and Svendsen, N. G. (2011). Evaluation of compost blankets for erosion control from disturbed lands, *J. Environ. Manage.* 92(3): 803-812.

Blanco-Canqui, H. (2017). Biochar and soil physical properties, *Soil Sci. Soc. Am. J.* 81(4): 687-711.

Brtnicky, M., Datta, R., Holatko, J., Bielska, G., Gusiatin, Z. M., Kucerik, J., Hammerschmiedt, T., Danish, S., Radziemska, M., Mravcova, L., Fahad, S., Kintl, A., Sudoma, M., Ahmed, N. and Pecina, V. (2021). A critical review of the possible adverse effects of biochar in the soil environment, *Sci. Tot. Environ.* 796: 148756.

Cai, W., Huang, H., Chen, P., Huang, X., Gaurav, X., Pan, Z. and Lin, P. (2020). Effects of biochar from invasive weed on soil erosion under varying compaction and slope conditions: Comprehensive study using flume experiments, *Biomass Conv, Bioref.* https://doi.org/10.1007/s13399-020-00943-3.

Caesar-Tonthat, T. C., Busscher, W. J., Novak, J. M., Gaskin, J. F. and Kim, Y. (2008). Effects of polyacrylamide and organic matter on microbes associated to soil aggregation of Norfolk loamy sand, *Appl. Soil Ecol.* 40(2): 240-249.

Carter, M. R. and Stewart, B. A. (1995). *Structure and Organic Matter Storage in Agricultural Soils*, Boca Raton, FL: CRC Press.

Chandra, S. and De, S. K. (1982). Effect of cattle manure on soil erosion by water, *Soil Sci.* 133(4): 228-231.

Flanagan, D. C., Norton, L. D. and Shainberg, I. (1997). Effect of water chemistry and soil amendments on a silt loam soil–Part II: Soil erosion, *Trans. ASAE* 40: 1555-1561.

Garcıa-Orenes, F., Guerrero, C., Mataix-Solera, J., Navarro-Pedreno, J., Gomez, I. and Mataix-Beneyto, J. (2005). Factors controlling the aggregate stability and bulk density in two different degraded soils amended with biosolids, *Soil Till. Res.* 82(1): 65-76.

Gardiner, D. T. and Sun, Q. (2002). Infiltration of wastewater and simulated rainwater as affected by polyacrylamide, *J. American Water Resour. Assoc.* 38(4): 1061-1067.

Gholami, L., Karimi, N. and Kavian, A. (2019). Soil and water conservation using biochar and various soil moisture in laboratory conditions, *CATENA* 182: 104151.

Gilley, J. E. and Eghball, B. (1998). Runoff and erosion following field application of beef cattle manure and compost, *Biolog. Sys. Engineer. Papers and Publications* 18. https://digitalcommons.unl.edu/biosysengfacpub/18.

Gilley, J. E. and Risse, L. M. (2020). Runoff and soil loss as affected by the application of manure, *ASAE* 43: 1583-1588.

Goldberg, N., Nachshon, U., Argaman, E. and Ben-Hur, M. (2020). Short term effects of livestock manures on soil structure stability, runoff and soil erosion in semi-arid soils under simulated rainfall, *Geosciences* 10(6): 213. https://doi.org/10.3390/geosciences10060213.

Govers, G., Gimenez, R. and van Oost, K. (2007). Rill erosion: Exploring the relationships between experiments, modelling and field observations, *Earth Sci. Rev.* 84(3-4): 87-102.

Graber, E. R., Fine, P. and Levy, G. J. (2006). Soil stabilization in semi-arid and arid land agriculture, *J. Mater. Civ. Eng.* 18: 190-205.

Hargreaves, J. C., Adl, M. S. and Warman, P. R. (2008). A review of the use of composted municipal solid waste in agriculture, *Agric. Ecosys. Environmentalist* 123: 1-14.

Ilea, R. C. (2009). Intensive livestock farming: Global trends, increased environmental concerns, and ethical solutions, *J. Agric. Environ. Ethics* 22(2): 153-167.

Illera, V., Cala, V., Walter, I. and Cuevas, G. (1999). Biosolid and municipal solid waste effects on physical and chemical properties of a degraded soil, *Agrochimica* 43: 178-186.

Keren, R. and Shainberg, I. (1981). Effect of dissolution rate on the efficiency of industrial and mined gypsum in improving infiltration in a sodic soil, *Soil Sci. Soc. Am. J.* 45(1): 103-107.

Kimaro, D. N., Poesen, J., Msanya, B. M. and Deckers, J. A. (2008). Magnitude of soil erosion on the northern slope of the Uluguru Mountains, Tanzania: Interrill and rill erosion, *CATENA* 75(1): 38-44.

Lebron, I., Suarez, D. L. and Yoshida, T. (2002). Gypsum effect on the aggregate size and geometry of three sodic soils under reclamation, *Soil Sci. Soc. Am. J.* 66(1): 92-98.

Lee, S. S., Gantzer, C. J., Thompson, A. L. and Anderson, S. H. (2010). Polyacrylamide and gypsum amendments for erosion and runoff control on two soil series, *J. Soil Water Conserv.* 65(4): 233-242.

Lee, S. S., Shah, H. S., Awad, Y. M., Kumar, S. and Ok, Y. S. (2015). Synergy effects of biochar and polyacrylamide on plants growth and soil erosion control, *Environ. Earth Sci.* 74: 2463-2473.

Lee, C. H., Wang, C. C., Lin, H. H., Lee, S. S., Tsang, D. C. W., Jien, S. H. and Ok, Y. S. (2018). In-situ biochar application conserves nutrients while simultaneously mitigating runoff and erosion of an Fe-oxide-enriched tropical soil, *Sci. Total Environ.* 619-620: 665-671.

Lentz, R. D. and Sojka, R. E. (1994). Field results using polyacrylamide to manage furrow erosion and infiltration, *Soil Sci.* 158(4): 274-282.

Lentz, R. D. and Sojka, R. E. (2000). Applying polymers to irrigation water: Evaluating strategies for furrow erosion control, *Trans. ASAE* 43: 1561-1568.

Levy, G. J. (2012). Sodicity. In Huang, M. P., Li, Y. and Sumner, M. E. (Eds.), *Handbook of Soil Sciences: Resource Management and Environmental Impacts* (2nd edn.). Boca Raton, FL: CRC Press, 18-1-18-28.

Levy, G. J. and Mamedov, A. I. (2013). Soil susceptibility to deformation in different agricultural management practices: Assessment from water retention curve characteristics at low suction. *Catena Verlag, Advances in GeoEcology* 42: 129-147.

Levy, G. J. and Nachshon, U. (2022). Sodic soils. In *Encyclopedia of Soils in the Environment* (2nd edn.). Elsevier Ltd. https://doi.org/10.1016/B978-0-12-822974-3.00030-6.

Levy, G. J. and Warrington, D. N. (2015). Polyacrylamide addition to soils: Impacts on soil structure and stability. In Cirillo, G., Spizzirri, U. G. and Iemma, I. (Eds.), *Functional Polymers in Food Science: From Technology to Biology, Volume 2: Food Processing*. John Wiley and Sons, 9-32.

Levy, G. J., Levin, J., Gal, M., Ben-Hur, M. and Shainberg, I. (1992). Polymers' effects on infiltration and soil erosion during consecutive simulated sprinkler irrigations, *Soil Sci. Soc. Am. J.* 56(3): 902-907.

Loveday, J. (1976). Relative significance of electrolyte and cation ex-change effects when gypsum is applied to a sodic clay soil, *Aust. J. Soil Res.* 14(3): 361-371.

Mamedov, A. I., Beckmann, S., Huang, C. and Levy, G. J. (2007). Aggregate stability as affected by polyacrylamide molecular weight, soil texture and water quality, *Soil Sci. Soc. Am. J.* 71(6): 1909-1918.

Mamedov, A. I., Shainberg, I., Wagner, L. E., Warrington, D. N. and Levy, G. J. (2009). Infiltration and erosion in soils treated with dry PAM of two molecular weights and phosphogypsum, *Aust. J. Soil Res.* 47(8): 788-795.

Mamedov, A. I., Wagner, L. E., Huang, C., Norton, L. D. and Levy, G. J. (2010). Polyacrylamide effects on aggregate and structure stability of soils with different clay mineralogy, *Soil Sci. Soc. Am. J.* 74(5): 1720-1732.

Mamedov, A. I., Huang, C., Aliev, F. A. and Levy, G. J. (2017). Aggregate stability and water retention near saturation characteristics as affected by soil texture, aggregate size and polyacrylamide application, *Land Degrad. Develp.* 28(2): 543-552. https://doi.org/10.1002/ldr.2509.

Mamedov, A. I., Tsunekawa, A., Tsubo, M., Fujimaki, H., Ekberli, I., Seker, C., Ozturk, H. S., Cerdà, A. and Levy, G. J. (2020). Structure stability of cultivated soils from semi-arid region: Comparing the effects of land use and anionic polyacrylamide application, *Agronomy* 10(12): 2010. https://doi.org/10.3390/agronomy10122010.

Mamedov, A. I., Fujimaki, H., Tsunekawa, A., Tsubo, M. and Levy, G. J. (2021a). Structure stability of acidic luvisols: Effects of tillage type and exogenous additives, *Soil Till. Res.* 206. https://doi.org/10.1016/j.still.2020.104832.

Mamedov, A. I., Tsunekawa, A., Haregeweyn, N., Tsubo, M., Fujimaki, H., Kawai, T. K., Kebede, B., Mulualem, T., Abebe, G., Wubet, A. and Levy, G. J. (2021b). Soil structure stability under different land uses in association with polyacrylamide effects, *Sustainability* 13(3): 1407. https://doi.org/10.3390/su13031407.

Martinez, J., Dabert, P., Barrington, S. and Burton, C. (2009). Livestock waste treatment systems for environmental quality, food safety, and sustainability, *Bioresour. Technol.* 100(22): 5527–5536.

Martínez-Blanco, J., Lazcano, C., Christensen, T. H., Muñoz, P., Rieradevall, J., Møller, J., Antón, A. and Boldrin, A. (2013). Compost benefits for agriculture evaluated by life cycle assessment: A review, *Agron. Sustain. Dev.* 33(4): 721–732. https://doi.org/10.1007/s13593-013-0148-7.

Moffet, C. A., Zartman, R. E., Wester, D. B. and Sosebee, R. E. (2005). Surface biosolids application: Effects on infiltration, erosion, and soil organic carbon in Chihuahuan Desert grasslands and shrublands, *J. Environ. Qual.* 34(1): 299–311.

Morin, J., Goldberg, D. and Seginer, I. (1967). A rainfall simulator with a rotating disc, *Trans. ASAE* 10: 74–77.

Nadler, A., Malik, M. and Letey, J. (1992). Desorption of polyacrylamide and polysaccharide polymers from soil materials, *Soil Technol.* 5(1): 91–95.

Norton, L. D., Shainberg, I. and King, K. W. (1993). Utilization of gypsiferous amendments to reduce surface sealing in some humid soils of the eastern USA, *Catena Supplement* No. 24, Soil Sealing and Crusting, J. W. E. Poesen and M. A. Nearing, eds., Catena, Cremlingen, Germany.

Norton, L. D., Mamedov, A. I., Huang, C.-H. and Levy, G. J. (2006). Soil aggregate stability as affected by long-term tillage and clay type, *Catena Verlag, Advances in GeoEcology* 38: 422–429.

NRCS. (2015). Conservation practice standard amending soil properties with Gypsum Products. Proposed full text for practice standard code (vol. 333). https://www.nrcs.usda.gov/wps/PA_NRCSConsumption/download?cid=nrcsepr d340736&ext=docx [Accessed August 6, 2017].

Orts, W. J., Sojka, R. E. and Glenn, G. M. (2002). Polymer additives in irrigation water to reduce erosion and better manage water infiltration, *Agro Food Ind. Hi Tech* July/August: 37–41.

Parr, J. F. and Hornick, S. B. (1992). Agricultural use of organic amendments: A historical perspective, *Am. J. Altern. Agric.* 7(4): 181–189.

Paulin, B. and O'Malley, P. (2008). *Compost Production and Use in Horticulture*. Department of Primary Industries and Regional Development, Western Australia, Perth. Bulletin 4746.

Peng, X., Zhu, Q. H., Xie, Z. B., Darboux, F. and Holden, N. M. (2016). The impact of manure, straw and biochar amendments on aggregation and erosion in a hillslope ultisol, *CATENA* 138: 30–37.

Perucci, P. (1990). Effect of addition of municipal solid-waste compost on microbial biomass and enzyme activities in soil, *Biol. Fertil. Soils* 10(3): 221–226.

Peterson, J. R., Flanagan, D. C. and Tishmack, J. K. (2002). Polyacrylamide and gypsiferous material effects on runoff and erosion under simulated rainfall, *Trans. ASAE* 45: 1011–1019.

Plante, A. and Voroney, R. P. (1990). Decomposition of land applied oily waste and associated changes in soil aggregate stability, *J. Environ. Qual.* 27: 395–402.

Poesen, J. W. A. (1987). *The Role of Slope Angle in Surface Seal Formation, International Geography, Part II, V. Gardiner*, ed. New York: Wiley.

Qin, L., He, Z. L. and Stoffella, P. J. (2012). Land application of biosolids in the USA: A review, *Appl. Environ. Soil Sci.* 2012: Article ID 201462. https://doi.org/10.1155/2012/201462.

Ramola, S., Belwal, T. and Srivastava, D. R. (2020). Thermochemical conversion of biomass waste-based biochar for environment remediation. In: Kharissova, O., Martínez, L. and Kharisov, B. (Eds.), *Handbook of Nanomaterials and Nanocomposites for Energy and Environmental Applications*. Cham: Springer. https://doi.org/10.1007/978-3-030-11155-7_122-2.

Sadeghi, S. H., Kiani-Harchegani, M., Hazbavi, Z., Sadeghi, P., Angolu-Jaramillo, R., Lassabatere, L. and Younesi, H. (2020). Field measurement of effects of individual and combined application of biochar and polyacrylamide on erosion variables in loess and marl soils, *Sci. Total Environ.* 728: 138866.

Seybold, C. A. (1994). Polyacrylamide review: Soil conditioning and environmental fate, *Commun. Soil Sci. Plant Anal.* 25(11–12): 2171–2185.

Shainberg, I., Sumner, M. E., Miller, W. P., Farina, M. P. W., Pavan, M. A. and Fey, M. V. (1989). Use of gypsum on soils: A review, *Adv. Soil Sci.*, Vol. 9. New York: Springer, 2–101.

Smith, H. J. C., Levy, G. J. and Shainberg, I. (1990). Waterdrop energy and soil amendments: Effects on infiltration and erosion, *Soil Sci. Soc. Am. J.* 54(4): 1084–1087.

Sojka, R. E. and Surapaneni, A. (2000). Potential use of Polyacrylamide (PAM) in Australian agriculture to improve off- and on-site environ- mental impacts and infiltration management, A Report to the Australian Land and Water Resource Research Development Council, The Institute for Sustainable Irrigated Agriculture, Goulburn-Murray.

Sojka, R. E., Bjorneberg, D. L., Entry, J. A., Lentz, R. D. and Orts, W. J. (2007). Polyacrylamide in agriculture and environmental land management, *Adv. Agron.* 92: 75–162.

Tang, Z., Yu, J., Lei, T., Shainberg, I., Mamedov, A. M., Ben-Hur, M. and Levy, G. J. (2006). Runoff and erosion in sodic soils treated with dry PAM and phosphogypsum, *Soil Sci. Soc. Am. J.* 70(2): 679–690.

Tejada, M. and Gonzalez, J. L. (2007). Influence of organic amendments on soil structure and soil loss under simulated rain, *Soil Till. Res.* 93(1): 197–205.

Tisdall, J. M. and Oades, J. M. (1982). Organic matter and water stable aggregates in soils, *J. Soil Sci.* 33(2): 141–163.

Truman, C. C., Nuti, R. C., Truman, L. R. and Dean, J. D. (2010). Feasibility of using FGD gypsum to conserve water and reduce erosion from an agricultural soil in Georgia, *CATENA* 81(3): 234–239.

Wallace, A. (1998). Use of water-soluble polyacrylamide for control of furrow irrigation-induced soil erosion. In: Wallace, A. and Terry, E. R. (Eds.), *Handbook of Soil Conditioners*. New York: Marcel Dekker, 445–462.

Wani, I., Narde, S. R., Huang, X., Remya, N., Kushvaha, V. and Garg, A. (2021). Reviewing role of biochar in controlling soil erosion and considering future aspect of production using microwave pyrolysis process for the same, *Biomass Conver Biorefin.* https://doi.org/10.1007/s13399-021-02060-1.

Warrington, D., Shainberg, I., Agassi, M. and Morin, J. (1989). Slope and phosphogypsum effects on runoff and erosion, *Soil Sci. Soc. Am. J.* 53(4): 1201–1205.

Wei, L., Li, F., Cai, D., Liu, Y., Luo, D. and Wen, Y. (2023). Investigating the effect of biochar application on raindrop-driven soil erosion under laboratory rainfall experiments, *Geoderma* 430: 116291.

Whalen, J. K. and Chang, C. (2002). Microaggregate characteristics in cultivated soils after 25 annual manure applications, *Soil Sci. Soc. Am. J.* 66(5): 1637–1647.

Woodrow, J. E., Seiber, J. N. and Miller, G. C. (2008). Acrylamide release resulting from sunlight irradiation of aqueous polyacrylamide/iron mixtures, *J. Agric. Food Chem.* 56(8): 2773–2779.

Yu, J., Lei, T., Shainberg, I., Mamedov, A. I. and Levy, G. J. (2003). Infiltration and erosion in soils treated with dry PAM and gypsum, *Soil Sci. Soc. Am. J.* 67(2): 630–636.

Zhang, F., Huang, C., Yang, M., Zhang, J. and Shi, W. (2019). Rainfall simulation experiments indicate that biochar addition enhances erosion of loess-derived soils, *Land Degrad. Dev.* 30(18): 2272–2286.

Zhu, L.-D. and Hiltunen, E. (2016). Application of livestock waste compost to cultivate microalgae for bioproducts production: A feasible framework, *Renew. Sustain. Energy Rev.* 54: 1285–1290.

Index

www.ingramcontent.com/pod-product-compliance
Ingram Content Group UK Ltd.
Pitfield, Milton Keynes, MK11 3LW, UK
UKHW021826100926
8556IPUK00003B/186